AF587522

Experiments and Validated Models for Adsorption Thermal Energy Storage in Industrial and Residential Applications

Experimentelle Untersuchungen und validierte Modelle eines thermischen Adsorptionsspeichers für Industrie- und Gebäudeheizprozesse

Von der Fakultät für Maschinenwesen der
Rheinisch-Westfälischen Technischen Hochschule Aachen
zur Erlangung des akademischen Grades eines
Doktors der Ingenieurwissenschaften
genehmigte Dissertation

vorgelegt von

Heike Schreiber

Berichter: Univ.-Prof. Dr.-Ing. André Bardow
Univ.-Prof. Dr.-Ing. Gerhard Schmitz

Tag der mündlichen Prüfung: 12. Juli 2017

Diese Dissertation ist auf den Internetseiten der Universitätsbibliothek online verfügbar.

Aachener Beiträge zur Technischen Thermodynamik Band 10
Heike Schreiber

Experiments and Validated Models for Adsorption Thermal Energy Storage in Industrial and Residential Applications

Experimentelle Untersuchungen und validierte Modelle eines thermischen Adsorptionsspeichers für Industrie- und Gebäudeheizprozesse

ISBN: 978-3-95886-178-7

Bibliografische Information der Deutschen Bibliothek
Die Deutsche Bibliothek verzeichnet diese Publikation in der Deutschen Nationalbibliografie; detaillierte bibliografische Daten sind im Internet über http://dnb.ddb.de abrufbar.

Herstellung & Vertrieb:

1. Auflage 2017

Süsterfeldstr. 83, 52072 Aachen
Tel. 0241/87 34 34
Fax 0241/87 55 77
www.Verlag-Mainz.de

ISSN: 2198-4832

Satz: nach Druckvorlage des Autors
Umschlaggestaltung: Druckerei Mainz

printed in Germany
D82 (Diss. RWTH Aachen University, 2017)

Danksagung

Die vorliegende Arbeit entstand am Lehrstuhl für Technische Thermodynamik der RWTH Aachen im Rahmen meiner Tätigkeit als wissenschaftliche Mitarbeiterin. Meine Jahre am Lehrstuhl wurden geprägt von wichtigen Personen, und es ist nun an der Zeit ihnen dafür Danke zu sagen.

Meinen herzlichen Dank möchte ich Prof. Dr.-Ing. André Bardow aussprechen, für die gebotenen Möglichkeiten zu lernen und zu lehren, für die stets konstruktive Kritik, für die wertvollen Diskussionen zu fachlichen Inhalten und darüber hinaus. Meine Wertschätzung gilt dabei nicht zuletzt seiner Art den Lehrstuhl zu leiten, mich und meine Kollegen in unserer Entwicklung zu fördern und uns eigenverantwortliches Handeln zu ermöglichen und abzuverlangen.

Des Weiteren gilt mein Dank Prof. Dr.-Ing. Gerhard Schmitz für sein Interesse an dieser Arbeit und die Übernahme des Koreferats im Promotionsverfahren. Ich danke Prof. Dr.-Ing. Reinhold Kneer für vielfältige Möglichkeiten der Weiterbildung, fachlich wie persönlich, während meines Studiums und schließlich für die Übernahme des Vorsitzes der Prüfungskomission im Promotionsverfahren.

Allen meinen Kollegen am Lehrstuhl, gegenwärtigen wie ehemaligen, danke ich für das großartige Arbeitsumfeld und die Freundschaften, die daraus entstanden sind. Angefangen hat es mit der Diplomarbeit bei Birger Klitzing: ihm verdanke ich einen guten Start am Lehrstuhl. Dr.-Ing. Franz Lanzerath danke ich für die unterstützende Begleitung meiner gesamten Promotionszeit und für gute Zusammenarbeit. Ich danke Stefan Graf für das Engagement in seiner Masterarbeit, die der Startschuss zu meiner Dissertation war und für seine Hilfsbereitschaft, bis hin zum Korrekturlesen dieser Arbeit. Ebenso danke ich Uwe Bau, dessen Begeisterung für unsere Modelle ansteckend ist, für Motivation und Zusammenarbeit in gemeinsamen Projekten. Besonderer Dank gilt Meltem Erdoğan für stetige persönliche Unterstützung und Wegbegleitung. Jan Seiler danke ich für spannende Diskussionen und viele gemeinsame Stunden in der Lehre. Liebe SORPer, es hat Spaß gemacht mit euch! Des weiteren danke ich meinen Studenten: Andreas Gebhardt, Stephan Merkelbach, Daniel Becker, Julia Moll, Christoph Grüntgens, Christiane Reinert und Sophia Schröer haben mit ihrem Engagement bei der Bearbeitung Ihrer Projekte zu dieser Arbeit beigetragen.

Mein besonderer Dank gilt allen technisch-administrativen Mitarbeitern des Lehrstuhls: ohne euch geht es nicht! Hartmut de Vries, Thomas Bungert, Djuro Dragas, Bernhard Müller und Edgar Brauers danke ich für ihre Hilfsbereitschaft, die über das normale Maß weit hinaus geht. Ich danke Gregor Migas, dass er nicht irgendwann aufgelegt hat, wenn ich ihn mit zahlreichen Rechnerproblemen behelligt habe. Iris Wallraven danke ich für allzeit hilfreichen Rat und die Organisation aller nötigen Termine und Räume.

Ich danke meinen Freunden, die immer an mich geglaubt und mich stets motiviert haben, allen voran Sarah-Ann Nepaschinks. Sandra und Malte Hinsch danke ich zudem für Raum und Zeit beim Zusammenschreiben. Jutta Müller-Trapet danke ich für ihre Fürsorge und die finale Rechtschreibprüfung dieser Arbeit. Meinen Eltern gilt der größte Dank für ihre liebevolle Unterstützung, für die Zuversicht und für alles, was sie mir mitgegeben und vorgelebt haben. Die Warmherzigkeit und Zähigkeit meiner Mutter und die Begeisterungsfähigkeit meines Vaters sind mein Fundament. Zum guten Schluss danke ich von Herzen meinem Freund, Markus Müller-Trapet. Während meiner Promotionszeit hat er gleichzeitig hinter mir gestanden und war mein Vorbild. Seine Liebe und die Ruhe, die er ausstrahlt, geben mir Halt. Danke für alles!

Aachen, im August 2017 *Heike Schreiber*

‘Our greatest weakness lies in giving up.
The most certain way to succeed is always to try just one more time.’

Thomas Alva Edison

Contents

Kurzfassung

Thermische Energiespeicher leisten einen wesentlichen Beitrag zu einer nachhaltigen Energieversorgung: Sie ermöglichen die Flexibilisierung von Energiesystemen und die Einsparung von Primärenergie. Besondere Vorzüge haben Speichersysteme, die Adsorption zur Speicherung thermischer Energie nutzen: Sie bieten hohe Speicherdichten, können ein breites Temperaturspektrum bedienen und durch den Wärmepumpeneffekt Abwärmeströme in die Prozesswärmebereitstellung einbinden.

Die vorliegende Arbeit betrachtet daher einen thermischen Adsorptionsspeicher, der zur Raum- und Prozesswärmebereitstellung genutzt werden kann. Zur energieeffizienten Bereitstellung von Prozesswärme für einen diskontinuierlichen Industrieprozess kann der Adsorptionsspeicher mit einer Kraft-Wärme-Kopplungsanlage kombiniert werden. Diese Anwendung wird im Rahmen dieser Arbeit mittels dynamischer Simulationen untersucht: Zur Evaluierung des Betriebsverhaltens des Speichers wird ein mathematisches Modell erstellt und mit Hilfe von vorhandenen Messdaten parametriert. In einer Simulationsstudie werden Alternativtechnologien, wie Spitzenlastkessel und thermische Phasenwechselspeicher, zur Einbindung der Kraft-Wärme-Kopplungsanlage betrachtet und mit dem thermischen Adsorptionsspeicher hinsichtlich der erreichbaren Primärenergieeinsparung im Industrieprozess verglichen. Die Studie zeigt, dass durch Einsatz des Adsorptionspeichers bis zu 25 % Primärenergie eingespart werden kann. Dieses Einsparpotenzial wird erreicht, wenn beim Entladen ein Niedertemperaturwärmestrom aus dem Prozess vorhanden ist und beim Beladen Niedertemperaturwärme im Prozess benötigt wird.

Methodisch zeigt die Studie auf, dass genaue Modelle, die insbesondere die Wärmeverluste und die Dynamik des Speichers präzise abbilden, unumgänglich sind für eine sorgfältige Bewertung des Speichernutzens. Die Basis dafür wird in dieser Arbeit gelegt. Ein neuer Prüfstand ermöglicht die präzise Charakterisierung von thermischen Adsorptionsspeichern. In einer experimentellen Studie werden die Wärmeverluste des Adsorptionsspeichers analysiert. Anschließend werden Speicherzyklen von der Kurzzeitspeicherung bis hin zur saisonalen Speicherung von thermischer Energie vermessen und die Leistungsfähigkeit des Speichers analysiert. Der Speicher erreicht eine Wärmerückgewinnungsrate von 69–91 % und eine Energiespeicherdichte von 20.4 kW h/m^3.

Mit Hilfe weniger Wärmeverlustmessungen sowie einer Zyklusmessung wird das anfangs eingeführte Modell des Speichers verbessert und umfassend validiert. Das Modell erreicht eine – im Vergleich zur Literatur sehr gute – Vorhersagegenauigkeit für industrielle und Gebäudeheizanwendungen. Die vorliegende Arbeit bietet somit die Basis für zukünftige Systemanalysen und die Erschließung neuer Anwendungsfelder für thermische Adsorptionsspeicher.

Abstract

Thermal energy storage (TES) helps to reduce energy consumption and peak demands by balancing heat supply and demand on all time scales from short-term to seasonal. Thus, TES is an important technology to improve flexibility and efficiency of energy systems. In particular, adsorption TES systems, which exploit the enthalpy of adsorption, provide high energy storage density and high efficiency.

The present thesis therefore analyzes an adsorption TES unit for residential and industrial applications. Industrial energy supply can be made more efficient by integrating waste heat into the process heat supply and by using energy-efficient technologies. Adsorption TES contributes to both approaches: waste heat can be integrated via the heat pump effect and TES allows for energy-efficient cogeneration heat supply for batch processes.

We evaluate the energy efficiency of the heat supply for an industrial batch process by adsorption TES and cogeneration. To evaluate the performance, a dynamic model of an adsorption TES unit is developed. Measurements from earlier experimental investigations of an adsorption TES unit are used to calibrate the storage model. As benchmark, a peak boiler and TES based on a phase-change material are considered. Our comparison demonstrates the significance of the process conditions for the choice of the appropriate technology. The study shows that adsorption TES offers significant potential to increase the energy efficiency: primary energy consumption can be reduced by up to 25 %. The key is the availability of low-grade heat at times of discharging and of a low-grade heat demand when charging the storage unit.

The study reveals that a comprehensive evaluation of the storage performance requires dynamic models that precisely describe the storage performance and the heat losses in particular. The present thesis provides the basis with a new experimental setup to precisely characterize the adsorption TES unit. In an experimental analysis of the TES performance, we quantify the heat losses, the energy recovery ratio (69–91 %) and the energy storage density (20.4 $\mathrm{kW\,h/m^3}$) of the adsorption TES unit for varying charging temperatures and storage times ranging from continuous operation to seasonal storage.

The extensive experimental study provides the basis to improve our model of the adsorption TES unit. The model is calibrated to heat-loss measurements and a storage-cycle measurement. We quantify the simulation accuracy and validate the model with measurements at various process conditions. The model achieves a higher prediction accuracy than other models from literature. The thesis thus provides a basis for future investigations of energy systems to exploit the advantages of adsorption TES.

Notation

Latin letters

a	coefficient (J/kg)
A	adsorption potential (J/kg)
A	area (m^2)
c	specific heat capacity (J/(kg K))
d	thickness (m)
D	mass-transfer coefficient (m^2/s)
f	factor or function
$\mathcal{F}$	function
Gr	Grashof number (-)
h	height (m)
h	specific enthalpy (J/kg)
$\dot{H}$	enthalpy flow rate (J/s)
I	input variable
k_t	student's t-distribution factor (-)
m	mass (kg)
$\dot{m}$	mass flow rate (kg/s)
N	number of output variables
Nu	Nusselt number (-)
Pr	Prandtl number (-)
p	pressure (Pa)
P	parameter
$\dot{Q}$	heat flow rate (J/s)
Q	heat (J)
r	radius (m)
t	time (s)
T	temperature (K)
u	uncertainty
u	specific internal energy (J/kg)
U	internal energy (J)
$U^{95\%}$	expanded uncertainty
UA	heat-transfer coefficient (W/K)
V	volume (m^3)

$\dot{V}$	volume flow rate (m^3/s)
w	water loading (kg/kg)
W	volume of adsorbate (m^3/kg)
x	variable
X	differential state
Y	output variable
Z	algebraic state

Greek letters

α	heat-transfer coefficient ($W\,K/m^2$)
Δ	difference
ε	emissivity (-)
η	energy recovery ratio (-), efficiency (-), dynamic viscosity (Pa s)
λ	thermal conductivity (W/(K m))
ϱ	density (kg/m^3)
σ	Stefan-Boltzmann constant
τ	storage time (s)

Acronyms

A	adsorber
C+	heat from condensation is used
C-	heat from condensation is not used
CV	coefficient of performance
DAE	differential algebraic equations
ΔE_{rel}	relative deviation of thermal energy
E	evaporator
E+	heat is freely available for vaporization
E-	heat has to be provided for vaporization
ERR	energy recovery ratio
ESD	energy storage density
GUM	Guide to the expression of Uncertainty in Measurement
HTF	heat-transfer fluid
HX	heat exchanger
PEC	primary energy consumption
RMSD	root mean square deviation
TES	thermal energy storage
VDI	German association of engineers
VLE	vapor-liquid equilibrium

Subscripts and superscripts

0	start value
2 h	for 2 hours storage time
95%	confidence level of 95 %
ac	related to ads-cas
ad	adsorptive water (in adsorbed state)
ads	adsorber surface, adsorbent, adsorption (discharging)
ads-cas	adsorber to casing
A-E	adsorber to evaporator
A-env	adsorber casing to environment
A,HX	adsorbent to adsorber heat exchanger
AHX,i	element i of adsorber heat exchanger
A,in	adsorber inlet
A,out	adsorber outlet
A,lam	steel lamellae of the adsorber heat exchanger
calc	calculated
cas	casing
cas-env	casing to environment
ce	related to cas-env
circulation	oil circulation system
cond	condensation
conti	for continuous cyclic operation
cycle	storage-cycle measurements
des	desorption (charging)
ΔT	temperature difference
E-A	evaporator to adsorber
E,ave	average value for evaporator
E-env	evaporator to environment
E,HX	evaporator heat exchanger
EHX,i	element i of evaporator heat exchanger
E,i	inner side of evaporator heat exchanger tube
E,in	evaporator inlet
E,out	evaporator outlet
E,met	evaporator metal parts: steel trays and casing
eff	effective
evap	vaporization or evaporator
env	environment or ambient temperature
HX	heat exchanger
in	during charging
ins	insulation
L	Lanzerath

l	liquid
ln	logarithmic
loss	heat loss
lossless	for lossless operation
m	related to mass
max	maximum
meas	measured
min	minimum
norm	normalized
oil	heat-transfer oil
out	during discharging
p	at constant pressure
pipe-loss	heat loss of oil pipes
pl	related to pipe-loss
rel	relative
s	saturated
seasonal	seasonal storage
sim	simulated
steady-state	for steady state conditions
store	storage unit
storage	storage period
t	t-distribution
TC	thermocouple
v	vapor, vaporization
w	water
VDI	according to VDI
zeo	zeolite

1 Introduction

Climate change is an increasingly urgent problem that causes severe and irreversible impact for people and ecosystems [1]. Thus, climate change has to be stopped or at least be mitigated. To mitigate climate change, greenhouse-gas emissions have to be reduced drastically. The main measures to achieve this reduction are the decrease of fossil-energy consumption and the transformation of energy-supply systems towards more sustainable systems. These sustainable systems have to be based on renewable-energy sources. However, the increased use of renewable-energy sources requires a flexible energy system [2].

To provide flexibility to energy systems, energy storage is a promising option [3]. Energy storage helps to decouple supply and demand: surplus energy can be stored for later use and peak-energy consumption can be reduced. Energy storage thus helps to reduce energy consumption by improving the resource-use efficiency of an energy system [4].

Amongst the total energy consumption in the German energy system, up to 47 % of the energy is used to provide process heat and residential heating [5]. To efficiently provide this heat, a sustainable energy system can be supported by thermal energy storage (TES) technologies [6]. TES systems can be charged by both surplus heat and surplus electricity and hence effectively bridge the gap between energy supply and heat demand [7]. Further benefits of TES systems are, according to Hyman [6]: lower facility energy costs, reduced energy consumption, improved facility plant efficiencies, smaller energy production equipment requirements, and more flexible plant operations. Given their relevance, TES systems are in focus of this thesis.

In particular, TES systems based on sorption processes offer major benefits as recently reviewed by Yu et al. [8]. Sorption systems provide energy storage by separation of (at least) two materials. The stored energy can be recovered as a heat flow when the materials recombine and enthalpy of sorption is released. Due to the high enthalpy of sorption, very high energy storage densities can be achieved by sorption TES systems [9]. Sorption TES systems are further able to shift heat flows both in time and temperature: due to the heat pump effect, sorption TES can make use of waste heat for process heat supply.

Sorption systems can be divided into absorption and adsorption systems [8]. They differ by the physical state of the sorbent which is liquid in absorption and solid in adsorption systems. Absorption systems often suffer from crystallization and corrosion [10] and are thus more difficult to handle than adsorption systems, which are analyzed in this thesis.

Adsorption systems for thermal energy storage are studied by several research groups all over the world. In particular, advanced materials for adsorption TES are developed to provide even higher energy storage densities [11–15]. Yet, to efficiently exploit the benefits of advanced materials, a storage unit requires a compact design and has to provide an advantage to the overall system it is integrated in, e.g. to lead to more efficient energy use. However, even though many research studies are ongoing, very few efficient TES systems have been realized so far [16]. Further research on efficiencies and comprehensive analysis of system performance is required to harvest the potential of adsorption TES [8, 16]. Hence, this thesis focuses on the performance analysis of a given adsorption TES unit on the system level.

This thesis considers adsorption TES with a well-known material, zeolite, which is widely used as an eco-friendly catalyst [17] and detergent builder [18]. For adsorption TES, zeolite is mostly combined with water as eco-friendly adsorbate with high adsorption enthalpy. Zeolite thus offers particular benefits for adsorption TES [19], such as high energy storage densities. Moreover, adsorption TES with zeolite can be applied to different applications on various temperature levels [20]: both heating and refrigeration [21] may be provided.

Adsorption TES with zeolite and water can be operated in reactors which are open to the environment, or in closed reactors. For applications where compact and efficient storage devices are needed, Yu et al. [8] recommend closed adsorption systems. Such a closed adsorption TES unit is studied in this thesis.

Lass-Seyoum et al. [20] assume that adsorption units can store thermal energy without significant heat losses. This assumption holds true for the latent part of the stored thermal energy, but sensible thermal energy is partly lost during storage operation [22]. Since heat losses are crucial for TES performance, they need to be quantified to apply storage in practice.

Furthermore, a comprehensive performance analysis has to consider both the storage unit and the system integration. To analyze the storage performance in the system context, the interaction of the storage unit with the system has to be studied for all relevant operating conditions. Conducting such a study experimentally requires a lot of effort. A model-based approach facilitates the analysis, but requires a valid description of the characteristics of the storage system [23].

Besides the analysis of all relevant operating conditions, a thorough evaluation of the performance of an adsorption TES system requires a comparison to alternative solutions to provide a benefit to the energy system. These issues are addressed in this thesis which is structured as follows.

Structure of the thesis

This thesis contributes to characterizing the performance of adsorption TES systems through experiment and simulation. In Chapter 2, the current research and state of the art are briefly reviewed. We examine both the applications and the characterization methods of adsorption TES systems. Thereby, shortcomings of the existing methods are identified. In conclusion of Chapter 2, the particular contributions of this thesis are outlined.

In Chapter 3, we[1] briefly present the adsorption TES prototype that is used in this thesis.

The performance of adsorption TES is evaluated for a new application in Chapter 4. Based on preliminary experiments with our prototype, a dynamic model is developed to analyze adsorption TES for cogeneration heat supply of an industrial batch process. For this application, the potential of adsorption TES is compared to alternative solutions to reduce primary energy consumption.

The study in Chapter 4 reveals the need for the precise knowledge of storage performance including heat losses. Thus, in the following chapters, we elaborate the adsorption TES characteristics and their precise description.

To rigorously quantify the performance of our adsorption TES unit, a new experimental setup is developed and briefly described in Chapter 5.

In Chapter 6, we determine the heat losses of our adsorption TES unit from steady-state measurements of the adsorber. We further analyze the storage performance during cycle measurements for charging temperatures of 175–250 °C. The energy recovery ratio and the energy storage density are quantified for cycles with various storage times, ranging from direct discharging to seasonal storage. The experimental results allow for a refined modeling of our adsorption TES unit.

[1]This thesis is written in the *pluralis modestiae* to avoid the excessive use of passive voice.

Chapter 7 contains a description of the model improvements and of the calibration approach. The model is further validated with additional measurements covering various operation conditions to enable a simulation-based analysis of storage applications.

We finally draw the conclusions to this thesis in Chapter 8 and outline ideas for further research.

2 Current Applications and Characterization Methods of Adsorption Thermal Energy Storage

In this chapter, we present the state of the art that is relevant to this thesis. The adsorption process is introduced for thermal energy storage in Section 2.1. Then, we identify the steps required to analyze the performance of adsorption thermal energy storage (adsorption TES) with the future goal to successfully apply adsorption TES systems in practice. Measures for the performance of an adsorption TES system are introduced in Section 2.2.

In Section 2.3, an overview of applications that benefit from the use of adsorption TES systems is given. The next sections deal with the characterization and modeling of adsorption TES units: in Section 2.4, we examine the influence of heat losses on adsorption TES performance and the methods to quantify heat losses. Finally, the state of the art in modeling of adsorption TES systems is presented in Section 2.5. In conclusion of this chapter, the contribution of this thesis is outlined in Section 2.6.

2.1 The principle of adsorption thermal energy storage

Thermal energy storage (TES) technologies are classified by their physical principle to store energy: sensible, latent or thermochemical [24]. Sensible TES is the storage of thermal energy by increasing the temperature of a storage material. Latent TES systems make use of the large enthalpy change during phase transition of a storage material. Thermochemical energy storage refers to reversible thermochemical reactions, absorption or adsorption processes for thermal energy storage [25]. For a general overview of TES technologies and applications, we refer to Cabeza et al. [24]. Kerskes [25] gives an overview of the topic of thermochemical energy storage with some practical examples.

Thermochemical energy storage materials usually consist of two components (= working pair) that are separated during charging, usually with one component in gaseous state. The enthalpy of reaction or of sorption of the working pair provides the basis for the TES performance. Amongst the storage principles, thermochemical energy storage has the greatest potential for high storage capacities and energy storage for long periods of time with limited heat loss [16]. A recent review on thermochemical energy storage systems can be found in the PhD-thesis of Fopah Lele [26].

The required temperature range of storage use determines which storage material is suited. Yan et al. [27] review available thermochemical reaction materials for thermal energy storage. Thermochemical reactions are primarily applied to storage at very high temperatures (300–1000 °C) such as storage in concentrated solar power plants [28], whereas sorption materials are rather applicable to middle and low temperature storage (20–300 °C), such as storage of solar thermal energy. Yu et al. [8] published an overview of sorption TES for solar energy.

The basic working principle is similar for all types of thermochemical energy storage. Thermo-chemical energy storage systems can be operated in closed or open reactors. The open-reactor mode is usually operated with ambient air and with water as the gaseous component of the working pair. Closed-reactor systems can work with all kinds of working pairs and at high pressure or vacuum conditions. Figure 2.1 shows the working principle for the TES system investigated in this thesis: a closed reactor with the adsorption working pair zeolite and water. Zeolite is then called the adsorbent and water is the adsorptive.

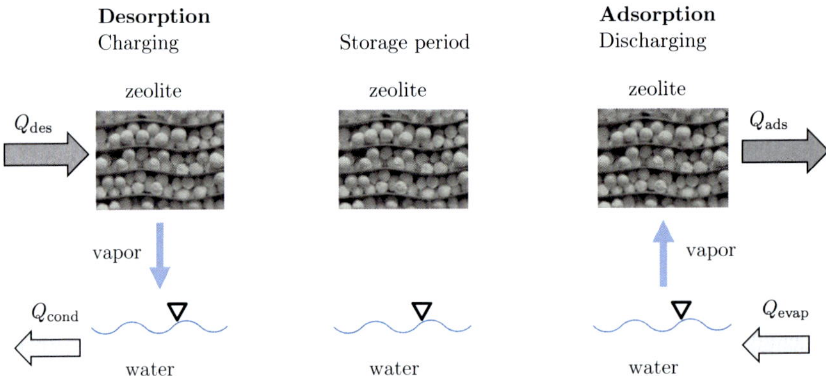

Figure 2.1: Working principle of adsorption thermal energy storage with zeolite and water: during charging, the heat input Q_{des} is used to desorb the zeolite. Released water vapor is condensed. During discharging, water evaporates and is adsorbed by the zeolite, releasing the heat output Q_{ads}.

During charging, the heat input Q_{des} leads to an increase of the temperature in the zeolite and to desorption, i.e. to the release of water vapor. The released water vapor can be condensed at a lower temperature. The enthalpy of condensation can be used as low temperature heat Q_{cond} or be dismissed to the environment. During the storage period, zeolite and water are stored separately until the stored thermal energy is needed. During discharging of the storage unit, the charging process is reversed: low-grade heat Q_{evap} is used to evaporate water at low temperatures. The water vapor is adsorbed by the zeolite. Hereby, the enthalpy of adsorption is released and can be used as process heat Q_{ads} at a higher temperature.

Closed-adsorption TES systems thus have the same operation modes as adsorption heat pumps [29]. In contrast, adsorption and desorption periods of the storage process are often separated in time. The details of the storage process, including temperatures and period times, are determined by the energy system in which the TES unit is integrated.

If the energy system faces a mismatch between thermal energy supply and demand, a TES unit may provide the necessary flexibility to overcome this mismatch. The mismatch can be in temperature, heat flow rate, space or time. The temporal mismatch defines the storage time, i.e. long-term or short-term storage. Spatial mismatch requires mobile storage systems. A mismatch in heat flow rate is overcome by thermal-load shifting. A mismatch in temperature can be met by heat transformation, i.e. by using the heat-pump effect. Adsorption TES is able to provide all kinds of this flexibility to thermal energy systems.

To provide flexibility, the TES system has to provide the required thermal power during discharging. Controlling the power output is an unsolved issue for adsorption TES systems [30], due to the pronounced adsorption dynamics. For an open-adsorption TES system, Michel et al. [31] observe that the inlet-air conditions, such as the humidity, are relevant to regulate the thermal power and the temperature of the heat output. The equivalent to the input-air humidity of an open system would be the water vapor pressure in a closed system. For a closed system, Yu et al. [30] propose to control the power output by regulating the flow rate of the heat transfer fluid with a variable frequency pump. However, research on this control strategy is still due.

2.2 Assessment of adsorption thermal energy storage

In general, a thermal energy storage (TES) system should provide a benefit to an energy system, such as lower energy consumption or smaller energy-supply-equipment requirements [6] compared to alternative solutions. To achieve a benefit, a TES unit has to provide unique features, such as heating and cooling with the same unit or it has to be compact and offer good performance, i.e. high energy efficiency.

The compactness of a TES unit can be measured with the energy storage density ESD. The energy storage density ESD is generally defined as the ratio of the discharging heat Q_{used} and a reference value that describes the size of the system: typically, the reference is a mass m_{ref} or a reference volume V_{ref}:

$$\text{ESD} = \frac{Q_{\text{used}}}{V_{\text{ref}}} \quad \text{or} \tag{2.1}$$

$$\text{ESD}_{\text{m}} = \frac{Q_{\text{used}}}{m_{\text{ref}}} \; . \tag{2.2}$$

Some studies calculate the energy storage density ESD_{m} based on the mass of the dry adsorbent [32, 33]. Typical adsorbents though have a low volumetric mass density, which certainly leads to larger values of the mass-related ESD_{m} than for volume-related definitions.

The definition of the reference volume ranges from the volume of the adsorbent material [29, 34], the volume of the adsorbent plus the heat exchanger, the volume of the adsorber including heat exchanger and casing [35], to the volume of the complete storage unit including adsorber and evaporator [29]. The reference value is crucial when it comes to comparing performance measures of storage systems.

Besides the reference value, the definition of the ESD may also vary in the energy used for the calculation. Aydin et al. [16], for example, calculate the ESD with the maximum chemical potential of the material. The maximum chemical potential of an adsorption material can be calculated by multiplying the maximum amount of adsorptive in equilibrium with the adsorbent, with the enthalpy of adsorption. This definition of the ESD is thus independent from any application or temperatures. However, the achievable energy storage density ESD strongly depends on the process conditions and the profile of the heat demand and supply [8], i.e. the temperatures of adsorption, desorption, condensation, vaporization, the time profile and the power requirements.

The performance of a TES system can be quantified with the energy recovery ratio ERR. In general, the energy recovery ratio ERR is defined as

$$\text{ERR} = \frac{Q_{\text{used}}}{Q_{\text{effort}}}, \tag{2.3}$$

i.e. the ratio of the usable heat Q_{used} released from the storage unit and used in the process to the heat Q_{effort} provided to the storage unit.

The heat flows of the TES unit can be transferred at various temperatures. These temperatures have a strong influence on the performance. For an adsorption TES system, the low-grade heat, which is transferred at low temperature during condensation and vaporization, is crucial to the performance. However, the effect of the low-grade-heat temperature on storage performance is still largely unexplored [8].

The temperatures of the heat flows of the TES unit are determined by the energy system. Besides temperatures, the heat flow rates are also influenced by the energy system. In general, thermal energy storage is not a stand-alone technology, but operates in response to the system [6]. As a result, the entire energy system has to be considered to quantify the storage potential for energy savings.

Besides energy savings, the economics of TES systems are relevant. Economic evaluation is implicitly considered, because performance measures used in this thesis are related to the economics of a TES system. The economics of a TES system depend on three factors [36]: (1) the costs per installed capacity, related to the energy storage density ESD; (2) the energy costs, related to the energy recovery ratio ERR; and (3) the number of cycles per year, which is determined by the application.

Finally, to get a definite statement on the benefits of adsorption TES, we need to benchmark adsorption TES to the most energy efficient supply system for a specific application. Depending on the process, the most efficient energy supply can be achieved by different technologies [37]. Typically however, adsorption TES concepts are not compared to the best alternative solution, but to solutions without any storage [38], or to water storage [39].

According to Aydin et al. [16], the goal of obtaining competitive adsorption TES systems requires more research on efficiencies and comprehensive performance analysis. We can conclude from this section that a holistic assessment of adsorption TES requires (1) the analysis of the storage performance depending on the process conditions and (2) the comparison to alternative technologies for a given application.

2.3 Applications for adsorption thermal energy storage

Adsorption systems can store thermal energy over a wide range of temperatures [19]. Therefore, adsorption thermal energy storage (TES) systems are suited for various applications; examples are given in the following literature review. Starting with commercialized adsorption TES systems in Section 2.3.1, Section 2.3.2 focuses on residential heating, which is widely considered for adsorption TES. Section 2.3.3 deals with TES for industrial processes.

2.3.1 Current commercial applications

Two known commercial domestic applications are small-scale adsorption TES systems with few kg of adsorbent material: an energy efficient dishwasher with an open-adsorption TES unit [40] and a self-cooling beer keg with a closed-adsorption TES unit [41]. These two example applications gain unique features by adsorption TES.

The dishwasher uses the water adsorption capacity of zeolite for drying the dishes. The integration of the adsorption TES unit into the dishwasher system allows exploiting both the adsorption enthalpy during discharging, i.e. the dish-drying period, and the condensation enthalpy of the desorbed water during charging, i.e. the dish-washing period. The heat for desorption, used to dry the zeolite, is thus partly recovered by condensation to heat the wash water. The smart TES integration saves up to 25 % energy compared to a conventional dish-washer [42].

The self-cooling beer keg makes use of the refrigeration capacity of adsorption systems. The beer keg is equipped with a jacket of water and preconditioned, dry zeolite, separated by a valve. Once the consumer opens this valve, the water is adsorbed by the zeolite and evaporative cooling yields cold beer. Hence, the adsorption TES unit provides the consumer with a cold beverage at any time, without a fridge or electricity at hand. The storage unit is recharged at 500 °C by the keg manufacturer while the beer is refilled [41].

2.3.2 Residential heating

The application of residential heating is characterized by temperatures for discharging between 30–80 °C [43] and the main use period is in winter. Many studies of adsorption TES deal with residential heating applications, because residential heating is considered promising to harvest a large energy savings potential [44]. Since this section focuses on adsorption systems, the reader is referred to Heier et al. [45] for an extensive literature review of thermal energy storage for residential heating.

For residential heating by solar heat, practical feasibility of adsorption TES has been demonstrated by Bales et al. [29], in the solar heating and cooling programme of the International Energy Agency (IEA SHC, task 32). The energy storage density of the investigated closed-adsorption TES system [46] was below the benchmark of a sensible water TES system [29]. As a result, research on enhanced storage materials was intensified [47].

Amongst others, Fopah Lele [26] is searching for new TES materials, based on thermochemical reactions, for residential applications. Yet, thermochemical storage materials suffer from low thermo-mechanical stability and slow reaction kinetics [47]. Li et al. [33] experimentally tested a closed system with a new adsorption material with enhanced water uptake. However, the design of the storage unit, a thin coating on a massive heat exchanger structure, prevents the storage unit from reaching high energy densities [48]. Yu et al. [32] characterized a closed, lab-scale reactor with a novel ad- and absorption cycle with a new material for thermal energy storage. Their prototype does not yet achieve the target energy storage density. Hence, Yu et al. [32] conclude that more research on system modeling and design is necessary.

An innovative system design for adsorption TES in residential heating was introduced by Hauer [41]: a district-heating net is equipped with a multipurpose, open-adsorption TES system. The TES unit is used once a day to provide heating in winter and cooling in summer. With 250 cycles per year and certain assumptions for thermal energy prices and investment costs, Hauer [41] estimate payback times of 8 years.

In the context of innovative systems for residential heating, Fehrenbach et al. [49] conclude that thermal energy storage can help to provide the flexibility which is necessary for load leveling in electricity grids. In combination with cogeneration or heat pumps, a TES unit can decouple electricity from heat supply. For the purpose of load leveling in electricity grids, adsorption TES can be applied to residential heating with storage periods of up to several hours.

Johannes et al. [50] experimentally investigate an open reactor with zeolite for 2-6 h energy release to be able to shave electricity peak loads in the French electricity grid. They successfully tested an 80 kg prototype with an energy recovery ratio of $\mathrm{ERR} = 34 - 55\,\%$. However, the energy recovery ratio has to be further improved to apply adsorption TES systems in practice.

Seasonal thermal energy storage

A special case of residential heating is much in focus of researchers: seasonal TES applications. For more information on seasonal TES in general, we refer to Xu et al. [51].

Mette et al. [52] give an overview of concepts for seasonal TES with thermochemical systems. Seasonal TES is currently studied on all scales: Dicaire et al. [53] investigate a small reactor with few grams of zeolite. Gaeini et al. [54] analyze system kinetics for an open reactor with 317 g of zeolite. Michel et al. [31] propose an open-bed prototype with 400 kg of a salt hydrate. On demonstration scale, two EU-projects develop closed-adsorption systems based on zeolite: Heatsaver-Demo [55] and COMTES [56].

Mostly, open adsorption reactors are applied as seasonal storage units, due to a simpler design [57]. Recently, Lefebvre and Tezel [58] published a review on adsorption TES for heating applications and conclude that adsorption TES can become a competitive technology for specific applications in the near future with material development and system optimization. But then, the economics are still very difficult for seasonal energy storage applications, due to long payback times with at most two cycles per year [36] (cf. Section 2.2).

2.3.3 Industrial processes

Considerable effort is invested to reduce energy expense and greenhouse gas emissions in industrial processes. In industrial processes, major energy savings can be obtained by establishing efficient energy supply systems [59] or by waste heat recovery.

In their recent review, Miró et al. [60] conclude a large potential for industrial waste heat recovery with thermal energy storage. However, only few thermochemical projects have been realized so far. One mobile TES system is reported: The MOPS prototype is a large-scale open-adsorption system based on 14 t of zeolite transporting heat from e.g. waste incineration to industrial drying processes [61, 62]. Krönauer et al. [62] conclude that the mobile storage system is nowadays not competitive to conventional fuels due to energy costs, but they are optimistic for future scenarios with CO_2-certificates trading.

Efficient energy supply can be realized by cogeneration providing combined heat and power [63]. However, combined supply of heat and power is only energy efficient if time profiles match for the different energy demands and supplies. A perfect match is rarely found in real processes. In this case, the heat demand usually dictates the operation of the cogeneration unit. But heat demand is often discontinuous, especially for batch processes, resulting in few full load hours of the cogeneration unit. To satisfy the discontinuous demand and to distribute heat, thermal energy storage can be applied [64]. Thermal energy storage then allows for smaller cogeneration units running at optimal efficiency and with increased full load hours [65].

In order to realize an efficient cogeneration system with thermal energy storage, the storage technology has to be chosen according to the specific process settings: first of all, the temperature of the heat supply from the TES unit has to fit the heat demand [66], which is in a temperature range of $T = 100 - 200\,°C$ for many industrial applications [67]. Adsorption TES offers a suitable large range of working temperatures between 60 and 300 °C [19].

Moreover, the energy storage density of the TES system should be high to reduce space and cost requirements. High thermal energy storage densities can be provided by adsorption [8]. In addition, adsorption TES allows exploiting the heat pump effect whereby low-grade heat can be integrated into the process heat supply to enhance energy efficiency [68]. Consequently, integrating adsorption TES into energy supply systems with cogeneration has the potential to reduce energy costs and greenhouse-gas emissions.

Closed adsorption systems are more compact and more efficient than open systems [8]. Closed adsorption reactors can even reach higher temperatures during discharging [69]. These features make closed-adsorption systems especially applicable for thermal energy storage in industrial processes.

Lass-Seyoum et al. [20] propose to combine cogeneration with a closed-adsorption system as thermal energy storage for temperature-flexible heat supply of industrial processes or heating applications. They test different reactor designs and experimentally determine storage densities on different reactor scales, for variable temperature conditions up to 200 °C. However, Lass-Seyoum et al. [20] did not specify an industrial process. The storage integration study presented in Chapter 4 is, to the best of our knowledge, the first to analyze the performance of a closed-adsorption TES unit in the context of a given industrial process.

Later, Shirai and Osaka [70] published a simulation study on industrial energy supply for food and automobile factories in Japan. They optimize the energy supply with a combination of cogeneration and a thermochemical energy storage system proposed by Kato et al. [38]. With an energy recovery ratio of 30 %, the TES system helps to reduce the primary energy use of the automobile factory by 1.5 %, but no potential for savings was found in the food factory [70].

A benchmarking study comparing two energy storage technologies was presented by De Boer et al. [71]. They compare a latent storage system to concrete storage for an industrial batch process to produce organic surfactants. The storage systems achieve energy savings of 50–70 %. De Boer et al. find out that the concrete storage offers higher efficiency increase, because it better matches the process requirements in terms of heat transfer rate and capacity, but with a very large reactor. Latent storage offers much higher storage densities compared to sensible TES, however, the latent storage system has to be severely oversized due to low heat transfer rates. Thus, the TES technology with the potentially higher energy storage density, i.e. latent TES compared to sensible TES [9], is not necessarily superior.

To conclude, industrial processes offer multiple applications for adsorption TES systems, mostly with higher temperatures than residential heating and cycle times of up to several hours [60]. Shorter cycle times increase the annual number of storage cycles, which is beneficial for the economic feasibility of a storage application [36]. However, higher temperatures lead to larger heat losses. The following section thus addresses heat losses of adsorption TES.

2.4 Heat losses of adsorption thermal energy storage

When thermal energy is stored, heat losses have to be taken into account [51]. Even though thermochemical energy storage systems are often said to store thermal energy without significant heat losses [20], this is only true for the enthalpy of adsorption. The thermochemical effect is lossless, but sensible thermal energy is lost during storage operation. Heat loss is often neglected, especially in preliminary design studies. However, a thorough evaluation of thermochemical energy storage systems requires consideration of heat losses during storage operation [22].

Xu et al. [51] emphasize that heat losses have to be addressed, in particular during discharging of adsorption thermal energy storage (adsorption TES) systems. After cooling down to ambient temperature, the storage materials have to be reheated to the discharging temperature, resulting in substantial heat losses, especially for seasonal TES systems [46]. Furthermore, Mette et al. [72] emphasize the importance of heat losses from the reactor wall to accurately describe the adsorption TES performance during charging. Dicaire et al. [34] attribute the difference between charging and discharging energy to both the reheating and to heat losses.

2.4.1 Influence of heat losses on storage performance

Losses of stored thermal energy have a negative impact on storage performance. Dawoud et al. [35] experimentally observe that the wall temperature of the storage casing decreases due to heat losses during a 5 h storage period. In particular, Dawoud et al. [35] attribute the heat losses of their closed-adsorption TES system to radiation inside the adsorber. They propose a radiation shield to reduce heat losses.

In experiments with a charging temperature of 150 °C and discharging to 40 °C, the system of Dawoud et al. [35] reaches a maximum energy recovery ratio of 67 % and an energy output density of 92 kW h/m^3 based on the volume of the adsorber.

Li et al. [33] experimentally investigate a closed-adsorption TES system with a coated heat exchanger at 70 °C charging temperature and discharging until 30 °C. Their system reaches an energy recovery ratio of 96 % and an energy storage density of 805 kJ/kg based on the adsorbent mass. To enhance the storage density, they propose to add more adsorbent between the heat exchanger fins. Li et al. [33] also propose a radiation shield to reduce heat losses.

Without heat losses, the energy storage density of adsorption TES systems increases with charging temperature [34], since more water is removed from the adsorbent increasing the storage capacity. But part of the thermal energy is stored sensibly. The ratio of sensible to latent energy[1] in an adsorption TES system depends on the working pair and on the temperature: e.g. Yu et al. [32] report a ratio of 0.75 (sensible/latent) at 85 °C charging temperature for their consolidated sorbent, i.e. activated carbon, silica solution and LiCl. For this system, the ratio decreases with charging temperature. In contrast, Jiang and Zhu et al. [73, 74] report a ratio of 0.9 (sensible/latent) at 155 °C, increasing with charging temperature, for their $MnCl_2$–$CaCl_2$–NH_3 resorption system. Besides the working pair and temperature conditions, the ratio of sensible to latent energy also depends on the geometry of the storage unit, i.e. the ratio of metal in the heat exchanger to adsorbent material. In any case, the part of the energy that is stored sensibly is prone to heat losses.

The heat losses certainly increase with temperature and lead to a decrease of the energy recovery ratio [34, 51]. To maximize the energy recovery ratio and the energy storage density, it is essential to determine the optimal storage temperature [51]. In order to determine a suitable storage temperature for optimal performance in terms of energy storage density and energy recovery ratio, knowledge of heat losses is important. Heat losses also have to be quantified for the determination of efficiencies during charging and discharging. These efficiencies are important measures to evaluate storage-cycle performance [75].

[1]In literature, this ratio is often referred to as "sensible to latent heat". In this thesis, we use the term "sensible to latent energy", which is thermodynamically more correct.

2.4.2 Determination of heat losses

Heat losses can be calculated on the basis of empirical Nusselt[2] correlations [76]. Heat-loss coefficients based on Nusselt correlations are used by Zanganeh et al. [77] in their simulation study of a large-scale sensible TES system; they find good agreement with measurement data. For a small-scale sensible TES system, Anderson et al. [78] show that heat losses based on Nusselt correlations enhance the simulation accuracy. Still, their simulations differ significantly from their measured results.

Nusselt correlations have also been used by Angrisani et al. [79] to calculate heat losses of a water storage unit. Their measured heat losses are four times larger than empirical predictions presumably due to thermal bridges. Gaeini et al. [54] experience the opposite: heat losses in the measurements are four times lower than calculated heat losses. Hence, heat-loss coefficients for TES units should be validated experimentally.

Schaube et al. [80] calculate a constant heat-loss coefficient from cooldown curves. Their calculations are based on strong simplifications to describe the change of internal energy of the storage unit. Bales et al. [81] use measurement data of charging and discharging cycles to fit a constant heat-loss coefficient for a closed-adsorption TES system. However, their result is influenced by the heat and mass transfer limitations inside the storage unit [82]. To decouple the processes inside from the heat losses outside of the storage unit, more information is necessary. Fopah Lele et al. [83] conclude that they cannot verify the heat losses of their closed thermochemical reactor, because of missing information on the reactor-wall temperature. Moreover, the heat flow rate on the reactor wall cannot be determined separately from transient measurements. To conclude, the simultaneous change of sensible and latent internal energy makes the quantification of heat-loss coefficients from storage-cycle or cooldown measurements difficult.

A steady-state approach to determine heat-loss coefficients for TES units was introduced by Mawire et al. [84]. Pal et al. [85] also use steady-state measurements to calculate heat losses for the open-adsorption storage system of Dicaire et al. [34]. From steady-state measurements of a latent cold-TES system, Osterman et al. [86] derive heat-loss coefficients and use them to estimate heat losses during storage-cycle measurements.

[2]Correlations for heat-transfer coefficients are named after Ernst Kraft Wilhelm Nußelt. In this thesis, we use the english spelling of his name: Nusselt.

2.5 Modeling of adsorption thermal energy storage units

Adsorption thermal energy storage (adsorption TES) is applicable to a variety of processes with different temperatures, heat flow rates and storage times. Thus, a comprehensive evaluation of adsorption TES storage performance has to consider a variety of operating conditions. A possible approach is to experimentally analyze the performance of a storage unit for various process conditions. Experiments like these can be done for specific process conditions, even for many process conditions, though with a lot of effort. And unfortunately, prototypes often do not achieve the expected storage capacity [23]. For a better understanding of the storage performance, modeling is essential [23]. We further need models to evaluate adsorption TES performance in the system context [50] and to be able to optimize the adsorption TES systems [87].

Storage performance cannot be analyzed without considering the interaction between the storage unit and the related system (cf. Section 2.3). Thus, the complete system needs to be simulated, which requires a well-chosen model complexity, balancing the trade-off between accuracy and simulation speed. According to Yong and Sumathy [88] three types of models are used to describe adsorption systems: thermodynamic steady-state models, dynamic lumped-parameter models, and models with detailed three-dimensional (3-D) description of heat and mass transfer phenomena. For more information on these spatially resolved models, Nagel et al. [89] provide an extensive review on models for thermochemical heat transformation and storage devices.

3-D models have a high model complexity and are hence not efficient for system analysis [90]. Steady-state models are also not useful for a system analysis, because adsorption is an intrinsically dynamic process. Dynamic models with accurate prediction of heat flow rates are thus necessary to capture the dynamics of the adsorption process [91].

2.5.1 Dynamic models with reduced complexity

Dynamic models with reduced complexity, e.g. lumped-parameter models, are suited for a system analysis with adsorption units [92]. Lanzerath [93] shows that lumped-parameter models of adsorbers with one-dimensional description of heat exchangers are accurate for modeling adsorption heat pumps and chillers. Closed-adsorption storage units have a similar design to heat pumps, with a heat exchanger surrounded by a packed bed of adsorbent. The packed bed can be modeled with a lumped model assuming homogeneous temperature and pressure distribution within the bed. The heat exchangers though have to be discretized in one dimension to capture rapid changes of temperature in the heat transfer fluids [93]. Hence, the lumped-model approach with discretized heat exchangers is also applied to closed-adsorption TES systems by Bales et al. [81].

Fernandes et al. [94] propose a lumped model with heat-transfer coefficients according to Nusselt correlations and with a contact resistance between the metal and the adsorbent according to Chua et al. [95]. Fernandes et al. [94] compare their model to a 2-D model by Brites [96] and show a deviation of 2 % in temperature and water uptake in the adsorber. Yet, a comparison to experimental data is needed to be sure that the model is accurate [94].

2.5.2 Comparison to experimental results

For closed-adsorption TES units, few models have been compared to experimental data: Fopah Lele et al. [83, 97] describe their closed thermochemical storage systems with 3-D models in COMSOL Multiphysics©. Comparison to experimental data reveals large differences between simulated and measured temperatures and pressure in the adsorption bed. Also the closed-adsorption TES models reported by Bales et al. [81] have partly been compared to experimental data of temperature and water content, showing qualitative agreement [29, 98].

In literature, we mostly find models of open adsorption TES systems [50, 54, 72, 80, 85, 99–101], which are packed beds without heat exchangers. The open systems with unidirectional mass flow can be accurately approximated by 1-D models [100] due to their simple geometries. In comparison, the geometries of closed-adsorption systems are more complex and are strongly simplified in the lumped model approach. These simplifications particularly influence the accuracy of heat and mass transfer models: the 3-D heat and mass transfer in packed beds with heat exchangers is described by effective coefficients that are empirical. Accurate heat and mass-transfer coefficients in the adsorber, evaporator and condenser are essential for high simulation accuracy [82]. Thus, reliable dynamic models for closed-adsorption TES systems can only be achieved by calibration to experimental data [102].

2.5.3 Calibration of empirical models

The MODESTORE adsorption system for heating applications [98] has been modeled using three heat-transfer coefficients that are calibrated to measurement data; mass-transfer limitations are not considered [81]. The simulation results fit reasonably well to the experiments: a visual comparison to the measured temperatures during discharging can be found in Reference [81], for temperatures during charging, see Reference [98].

Pal et al. [85] model the open storage system of Dicaire et al. [34] and calibrate their model parameters to experimental results by minimizing the error between measured and simulated temperatures. Nevertheless, their simulation result reveals systematic deviations to the measured temperatures and humidities. Tatsidjodoung et al. [101] adapt the mass transfer coefficient of an open reactor model by comparison to the measurements of Johannes et al. [50]. Tatsidjodoung et al. [101] conclude that the model can approximately predict the storage performance despite temperature deviations of 40 °C [101].

2.5.4 Definition of simulation accuracy

In most cases [54, 72, 80, 81, 83, 97–99, 101], model validation is interpreted as visual comparison of simulations to experimental results. These results mostly show a good qualitative agreement, but the agreement is rarely further quantified statistically: Tatsidjodoung et al. [101] quantify the deviations of simulated and measured temperatures with up to 30 %. Fopah Lele et al. [83] compare the temperatures and find a confidence interval of 82 % for charging and 70 % for discharging. Michel et al. [57] compare the measured and simulated water uptake in an open reactor and report an error of 6 % on the integrated value. Even if an accuracy is given for the simulated temperatures or water uptake, the accuracy of the model is still not comparable in terms of storage performance. The storage performance is measured in heat flow rates and energies and thus simulation accuracy should also be related to these quantities.

Gaeini et al. [54] compare simulated heat flow rates to experimental data for a discharging measurement without further commenting on the discrepancies. As there is little information on the accuracy of adsorption TES models to compare with, we take a look at models from adsorption heat pumps or chillers. For detailed information on modeling of adsorption chillers, we refer to a recent review by Pesaran et al. [103]. We find an accuracy[3] of performance indicators for adsorption chillers in Chen et al. [104]: their model describes the specific cooling power (SCP) with an accuracy of 5.3 % and the coefficient of performance (COP) with 6.5 %.

[3]The correct scientific term would be deviation, not accuracy. However, many authors use the term accuracy to have a positive expression for their measure of model quality, i.e. the accuracy is good for low deviations. In this thesis, we will stick to this nomenclature.

However, the accuracy of performance indicators such as the COP do not account for the system dynamics [93]. The accurate description of the dynamics is important for storage systems, in particular for load-shifting applications and short cycle times. To give a value for the accuracy of system dynamics, Lanzerath [93] defines the coefficient of variation CV based on the root mean square deviation $RMSD$ between a measured heat flow rate $\dot{Q}_{\text{meas}}$ and a simulated heat flow rate $\dot{Q}_{\text{sim}}$. With the duration of the measurement Δt_{meas}, the CV is defined as

$$CV = \frac{RMSD}{\left|\int \dot{Q}_{\text{meas}}\,\mathrm{d}t\right| / \Delta t_{\text{meas}}}, \text{ with} \tag{2.4}$$

$$RMSD = \sqrt{\frac{1}{\Delta t_{\text{meas}}} \int \left(\dot{Q}_{\text{meas}} - \dot{Q}_{\text{sim}}\right)^2 \mathrm{d}t}\;. \tag{2.5}$$

After thoroughly calibrating his adsorption chiller model, Lanzerath [93] achieves a simulation accuracy of $CV_{\text{evap}} = 9.1 - 20.2\,\%$ for the evaporator heat flow rate, $CV_{\text{cond}} = 16.8 - 33.3\,\%$ for the condenser heat flow rate and $CV_{\text{ads}} = 14.0 - 25.3\,\%$ for the adsorber heat flow rate. Lanzerath also reports an overall coefficient of variation for all heat flow rates of $CV = 13.3 - 21.7\,\%$.

However, chiller models usually neglect heat losses. This assumption is valid for adsorption chillers because of short cycle times and low temperatures, but it does not hold for storage applications (cf. Section 2.4). Modeling a storage unit without accounting for heat losses leads to large deviations between measured and simulated results [32, 101] (cf. Section 2.4).

In particular for applications with storage times, heat losses increase. Since the determination of heat losses is challenging (cf. Section 2.4.2), increased heat losses may lead to an increased uncertainty of the total energy, which is stored during charging or released during discharging. Belmonte et al. [105] define the simulation accuracy for total energies of their model of a latent TES unit: they give the relative deviation of the energy during charging and discharging as 10 % for a salt-hydrate system and approximately 30 % during charging of a storage unit with paraffin.

2.5.5 Model validation

The comparison of simulation results to experimental data is often restricted to a single measurement from a fixed application [81, 85, 98, 101]. However, the evaluation of storage performance requires simulating the storage behavior for different operating conditions.

Thus, the model should capture the dynamics of more than the calibration measurement. As Yong and Sumathy conclude, model validation is a key step in model development [88].

Lefebvre et al. [99] compare the simulated temperature and relative humidity of their open-reactor model to two measurements with different air flow rates and humidities. Schaube et al. [80] show plots of deviations between measured and simulated temperatures for various measurements of their open, high-temperature thermochemical TES system. Following the examples of Lefebvre and Schaube, the validation of each adsorption TES model should comprise comparison to various measurement conditions.

In contrast to the model of Schaube et al. [80], other models [32, 94, 97] are used to describe adsorption TES systems at charging temperatures below 100 °C. Yet, higher temperatures are required in many applications (cf. Section 2.3.3) and adsorption TES is flexible in temperature [20]. A valid model for an adsorption TES system should thus be able to describe storage performance for all the relevant working conditions. For this purpose, experimental validation of the model should also be performed at elevated temperatures.

For an open reactor with zeolite, Pal et al. [85] conclude that the charging performance of their system reaches an optimum at 250 °C. These elevated temperatures have to be provided by the experimental setup to allow model validation at elevated temperatures. Working temperatures up to 500 °C are easy to emulate in open-system experiments by applying simple air heaters [80]. However, for closed-adsorption systems, high temperatures are more difficult to provide due to high pressures in the heat-transfer-fluid circulation systems. As a result, most experimental facilities only provide heat at temperatures below 100 °C [32, 33, 46, 83, 98, 106]. Few experimental setups are able to provide higher temperatures to closed-adsorption TES units: Dawoud et al. [35] achieve 150 °C, Gantenbein reaches 180 °C [29] and Lass-Seyoum et al. [55] even reach 200 °C. However, models of closed-adsorption TES systems have not yet been validated, in particular not with experiments at temperatures above 100 °C.

2.5.6 Prediction accuracy

To evaluate the results of a simulation study, we want to know the accuracy of predicting storage performance for operating conditions other than the calibration condition. This prediction accuracy can be determined by simulating a storage cycle with operating conditions other than the calibration condition. The simulation result is further compared to a measurement of such a cycle. The agreement between the simulation and the measurement is then the prediction accuracy [93]. The quantification of the prediction accuracy makes the quality of the model measurable.

For an adsorption chiller, Lanzerath [93] determines the prediction accuracy for variations in temperatures and cycle times, i.e. the duration of de- and adsorption of their adsorption chiller, and for two different adsorbent materials: zeolite and silica gel. The total CV is 12.6–27.4 % for the silica gel-chiller operating at lower temperatures and 27.3–44.7 % for the zeolite-chiller at temperatures up to 160 °C. Lanzerath also calculates a prediction accuracy for the COP that is within the measurement uncertainty for the silica gel-chiller (7.2 %). In contrast, the COP is overestimated for the zeolite-chiller (18.7 %) due to neglecting heat losses.

2.6 Contribution of this thesis

The aim of this thesis is to contribute to exploiting the advantages of adsorption thermal energy storage (adsorption TES) for efficient future energy supply systems. The literature review (Chapter 2) reveals the challenges in evaluation and optimization of the adsorption TES technology: Storage performance has to be evaluated in the context of an appropriate application and should be compared to the best technology available (cf. Sections 2.2 and 2.3). Valid dynamic models of the storage behavior including heat losses are crucial to evaluate and optimize adsorption TES systems (cf. Sections 2.4 and 2.5). In the following, the particular contributions are highlighted:

Assessment of adsorption thermal energy storage for an industrial application This thesis provides a thorough evaluation of adsorption TES performance for heat supply with cogeneration in an industrial batch process. In a preliminary work, we experimentally demonstrated the feasibility of an adsorption TES prototype (Chapter 3) to support a cogeneration unit to supply heat to a brewing process [107]. Based on these experiments and a known modeling framework, we develop a dynamic model of the adsorption TES prototype and calibrate it to the measurement data (Chapter 4). With this model, we evaluate the storage performance for various process conditions (cf. Section 2.2), and resolve in particular the influence of the low temperature heat. To come to a conclusive evaluation (cf. Section 2.2), we further compare the adsorption TES system energetically to a conventional system and to another innovative energy storage technology. With this comprehensive assessment, the thesis provides the first integration study of adsorption TES in industrial process energy supply (cf. Section 2.3.3).

Assembly of a high temperature experimental setup The application study in Chapter 4 confirms the importance of an accurate description of the TES performance in general and of the heat losses in particular (see also Section 2.4). Adsorption TES performance has been experimentally investigated up to 200 °C so far (cf. Section 2.5.5). However, adsorption TES can be applied to processes with even higher charging temperatures (cf. Sections 2.3 and 2.5.5). Thus, an improved experimental setup is built to analyze TES units for applications with a large range of temperatures up to 260 °C (Chapter 5).

Analysis of heat losses and storage performance With the new experimental setup, the heat losses of our adsorption TES prototype are quantified by combining measurements of steady-state conditions (cf. Section 2.4.2) with measurements of the wall temperature of the adsorption TES unit. This approach even allows to estimate radiative heat losses inside the adsorber. Moreover, quantification of charging and discharging efficiencies (Section 2.4.1) is now possible, due to the determination of a valid heat-transfer coefficient for heat losses. Hence, with the thorough investigation of the heat losses, we can comprehensively analyze the performance of our adsorption TES prototype (Chapter 6). The analysis reveals the trade-off between energy recovery ratio and energy storage density for several temperatures and storage periods of residential heating applications.

Procedure to calibrate and validate dynamic adsorption TES models A useful model for adsorption TES systems accurately describes the storage performance for the full scope of possible applications (cf. Section 2.5). This thesis establishes a method to obtain such a valid dynamic model with a well-balanced complexity. In Chapter 7, we describe the thorough calibration of an improved model setup for our adsorption TES system.

Visual comparison of simulation and experiments is state of the art to determine the simulation accuracy of adsorption TES models (Section 2.5.4). In this thesis, we define and quantify the simulation accuracy to make the model quality comparable. Moreover, a comprehensive validation requires comparison to various measurement conditions, including elevated temperatures (cf. Section 2.5.5). We thus perform an exhaustive experimental validation of the model and are the first to quantify the prediction accuracy for an adsorption TES model (cf. Section 2.5.6). Finally, we discuss limitations of the used modeling approach. With the validated adsorption TES model, the thesis provides a sound basis for future research on new applications of adsorption TES systems.

3 A Small-Scale Adsorption Thermal Energy Storage Unit

The adsorption thermal energy storage (TES) unit in this thesis consists of two heat exchangers that have already been applied in previous studies at the Institute of Technical Thermodynamics of RWTH Aachen University [35, 107, 108]. In this chapter, we briefly describe the features of the adsorber heat exchanger (Section 3.1) and the evaporator/condenser heat exchanger (Section 3.2).

The heat exchangers were combined by Gasper as an adsorption heat pump [108]. Later, the system was introduced as a storage unit by Dawoud et al. [35]. In Section 3.3, we describe the two designs of the closed adsorption TES unit that were used in this thesis.

3.1 Adsorber

The stainless steel adsorber heat exchanger is made of a bundle of tubes connected by lamellae (Figure 3.1). The space between the lamellae is filled with the spherical adsorbent beads. A perforated sheet prevents the adsorbent from falling out of the adsorber.

The lamellae provide for a good heat transfer from the adsorbent beads to the heat transfer fluid. The fluid flow in the heat exchanger is serial, providing a sufficiently high velocity to ensure a turbulent flow regime even at low temperatures. Thus, the adsorber design is suitable for applications where high heat flow rates are needed. However, the large amount of metal reduces the space for active storage material: the zeolite to metal mass ratio of the adsorber heat exchanger is $m_{\text{zeolite}}/m_{\text{metal}} \approx 0.5$. The geometries of the adsorber and the evaporator are given in Table 3.1.

The adsorption working pair used in this thesis is zeolite 13 X as adsorbent and water as adsorptive. Water is thermally stable, cheap, non-toxic and provides for a high enthalpy of adsorption in combination with zeolites [109]. In the following, the water that is adsorbed by the zeolite will be referred to as adsorptive water, to distinguish between adsorptive and heat transfer fluid.

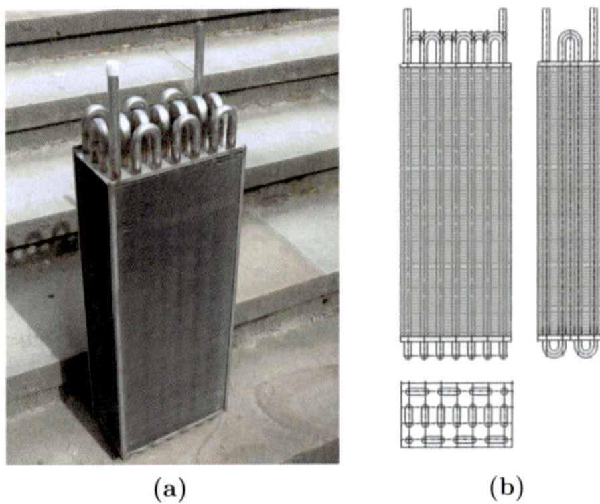

Figure 3.1: (a) Picture and (b) schematic of the adsorber heat exchanger with tubes and steel lamellae [108].

Table 3.1: Geometry of adsorber (cf. Figure 3.1) and evaporator (cf. Figure 3.2).

Adsorber		Evaporator	
mass of zeolite m_{zeolite}	10 kg	mass of water m_{w}	2.8 kg
mass of stainless steel m_{steel}	20.6 kg	mass of stainless steel m_{steel}	6.69 kg
number of tubes	28	number of tubes	3
length of each tube	0.6 m	length of each tube	3.55 m
outer diameter of tube	16 mm	outer diameter of tube	14.3 mm
number of lamellae	182		
thickness of lamellae	0.2 mm		
volume of adsorber V_{adsorber}	0.022 m^3	volume of evaporator $V_{\text{evaporator}}$	0.005 m^3

We choose zeolite 13 X and water, because the adsorption properties offer potential for high energy storage densities [19]. The equilibrium data of zeolite 13 X and water is reported in literature [110, 111]. For more information on the zeolite, see Appendix A.

3.2 Evaporator

The adsorptive water can both be evaporated and condensed in a single heat exchanger, which in the following is referred to as evaporator. The evaporator consists of three levels, each with a spiral, corrugated tube (Figure 3.2). The fluid flows through the three spiral tubes in parallel. Hence, the water flow is not necessarily distributed evenly to all three heat exchanger tubes, because the pressure drop may vary from tube to tube.

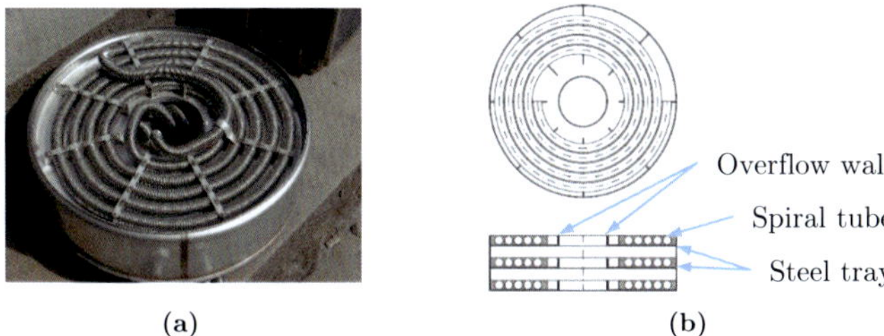

Figure 3.2: (a) Picture and (b) schematic of the evaporator heat exchanger with spiral, corrugated tubes on three levels [108].

The compact design of the evaporator offers a large heat exchanger surface for good vaporization and condensation characteristics. The adsorptive water inside the evaporator can condense on all three evaporator tubes and partly remains on steel trays below the spiral tubes. These steel trays are open to the center of the evaporator. On this inner side, the two upper-level steel trays have an overflow wall (cf. Figure 3.2). However, the overflow wall is lower than the tube diameter. Thus, the upper two levels are never completely flooded with water.

Underneath the lower spiral tube, the evaporator has a sump. Due to the sump, the lower level is still partly flooded with water when the upper two levels are already dry. When the upper levels fall dry, i.e. when the water filling level falls below a certain point, the heat transfer from the heat exchanger fluid to the adsorptive water strongly decreases. Still, there is water in the sump to evaporate, but at low flow rates.

3.3 Storage unit designs

In a preliminary study [107], we proposed to use an adsorption-TES unit for industrial heat supply of a brewing process. In an experimental study on the brewing-process application, we performed measurements with a TES unit containing the adsorber above the evaporator in one single container, cf. Figure 3.3 (a). The study in Chapter 4 is based on these measurements.

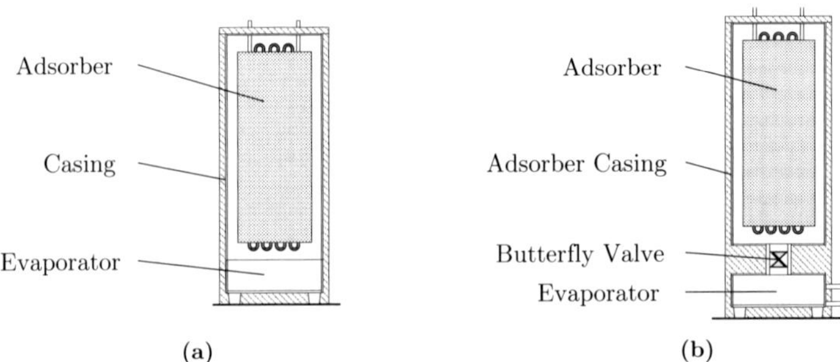

Figure 3.3: (a) Adsorption TES unit with adsorber and evaporator in one single container (Chapter 4). (b) Adsorption TES unit with a butterfly valve to separate adsorber and evaporator (Chapters 5–7).

The thesis further considers a configuration with a butterfly valve to separate adsorber and evaporator (cf. Figure 3.3, b) for expanded storage periods in Chapters 6 and 7.

With 10 kg of zeolite, the storage unit is small-scale compared to the 7 t adsorption-TES prototype of Hauer [41] or 200 kg in the adsorption-TES prototype of Jaehnig [46].

4 Adsorption Thermal Energy Storage for Cogeneration in Industrial Batch Processes

This chapter provides a comprehensive evaluation of an industrial application of adsorption thermal energy storage (adsorption TES). In a previous report [107], we propose to combine a cogeneration unit and an adsorption TES system for energy efficient heat supply of a brewing process. The brewing process is a representative example for an industrial batch process with time-varying heat demand on different temperature levels. In particular, small and medium sized breweries have large potential to improve their energy efficiency [112]. Sturm et al. [113] show that cogeneration can be profitably applied to meet the brewing process energy demand, including heat and electricity.

However, high investment costs can be an obstacle to apply cogeneration in practice. To make best use of the investment, the cogeneration unit should run at full load all the time. Full-load operation also allows to run the cogeneration unit with its highest efficiency [114]. Constant full-load operation is, however, not possible in batch processes with a time-varying heat demand. In this case, the cogeneration units have to be sized according to the peak demand leading to large investment costs while running at inefficient part-load conditions most of the time. Thus, reduced unit size with full-load operation is preferred, which is possible with the help of e.g. adsorption TES.

Our approach to assess adsorption TES for this application is to analyze the impact of adsorption TES on the primary energy consumption of the process. For this purpose, we develop a model of the adsorption TES unit from Chapter 3. By calibrating the model to measurement data from experimental investigations described in Section 4.1, we gain a reliable basis for our analysis (Section 4.2). The model is used to evaluate the integration

---

Contents of this chapter have been published in:

H. Schreiber, S. Graf, F. Lanzerath, and A. Bardow. "Adsorption thermal energy storage for cogeneration in industrial batch processes: Experiment, dynamic modeling and system analysis". *Applied Thermal Engineering* 89 (2015), pp. 485–493.

of cogeneration and adsorption TES into the brewing batch process (Section 4.3). In particular, we examine the effect of the temperature of the low-grade heat on the storage performance. In order to assess the true potential of adsorption TES for batch-process energy supply, we analyze different cogeneration integration concepts regarding their energy-savings potential (Section 4.4). For this purpose, we compare adsorption TES to alternative cogeneration integration concepts using either a conventional non-storage solution with a peak boiler or using innovative latent TES. Section 4.6 summarizes the results of our assessment.

4.1 Experimental emulation of the brewing process

Figure 4.1 illustrates the time profile of the brewing process that we analyzed in a previous study [107]. Once in every 6 hour brewery batch, a large amount of process heat is needed at a temperature of 120 °C for one hour. During the other 5 hours, the batch does not require process heat, i.e. heat at high temperature. The cogeneration unit then charges the adsorption TES unit with heat at up to 250 °C. This high temperature heat is provided by the exhaust-gas heat exchanger of the cogeneration unit. During discharging of the storage unit, the combined heat outputs of the storage unit and the cogeneration exhaust-gas satisfy the process-heat demand.

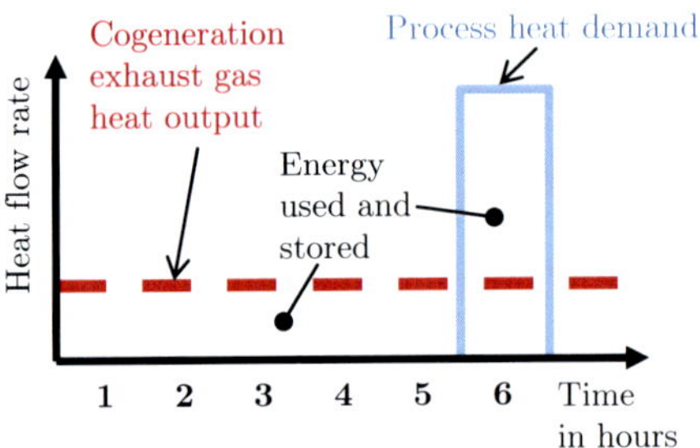

Figure 4.1: Brewing process-heat supply by cogeneration and storage: time profile of process-heat demand and supply.

The storage process of this brewery application was experimentally analyzed in a laboratory at the Institute of Technical Thermodynamics of RWTH Aachen University. The experiments were performed with the experimental setup shown in Figure 4.2. The setup contains two adsorption TES units that were connected in parallel. Each storage unit combines one adsorber and one evaporator into a vacuum-sealed container without a valve in between (cf. Figure 3.3 (a) in Chapter 3). No valve is needed between adsorber

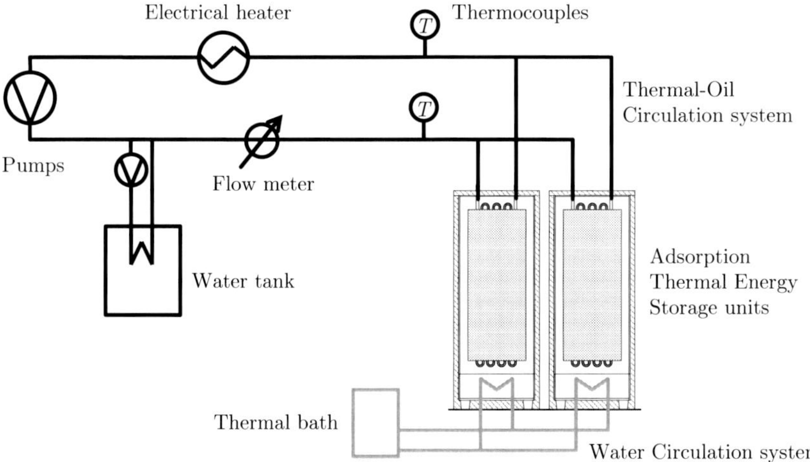

Figure 4.2: Scheme of the experimental setup where two adsorption TES units were tested. A thermal bath controls the temperature in the water circulation system. An electrical heater provides heat to the thermal oil. A water tank emulates the heat sink of the brewery batch.

and evaporator, because the charging and the discharging period immediately follow each other in the storage process of the brewery application.

In the experimental setup (Figure 4.2), the heat supply from the cogeneration unit was emulated in the laboratory using an electrical heater in a thermal-oil circulation system. The process-heat demand was emulated with a water tank, which represents the brewery batch. The temperature in the evaporator was controlled by a thermostat using water as heat-transfer fluid. Volume flows were measured with oval wheel meters. The volume flows could not be adjusted, as they were set by pumps in the thermal-oil circulation system and in the thermal bath. PT100 resistance thermometers of class A were used to determine the temperature differences between in- and outlet flows of the heat exchangers. Details on the measurement uncertainty of the experimental setup can be found in Appendix B.1.

During charging, the adsorbers of the two TES units were heated and desorbed with constant thermal output power of 2 kW from the electrical heater while the condenser temperature was kept constant. During discharging, the water tank was heated, representing the beer batch in the brewing process. To satisfy this heat demand, heat flows were combined from the adsorbers and the electrical heater. At the same time, the temperatures of both evaporators were kept constant.

The relevant heat flow rates $\dot{Q}$ were determined by measuring the flow rate $\dot{V}$ of the corresponding heat-transfer fluid (HTF) and the temperature difference ΔT across the corresponding heat exchanger (HX)

$$\dot{Q} = \left(\varrho c_{\mathrm{p}} \dot{V}\right)_{\mathrm{HTF}} (\Delta T)_{\mathrm{HX}} , \tag{4.1}$$

with the density ϱ and the heat capacity c_{p} of the heat-transfer fluid. Time-resolved measurements of the heat flow rates $\dot{Q}$ allow to quantify the dynamic performance of the adsorption TES unit, in particular the heat flow rates during charging and discharging. Measurements were conducted with constant electrical power. The temperature of desorption was up to 200 °C instead of 250 °C due to restrictions of the experimental setup. For adsorption, the temperature was 120 °C. The process-heat demand during discharging was fixed by the thermal capacity of the water tank. Temperatures for condensation and vaporization were varied, leading to three combinations $T_{\mathrm{cond}}/T_{\mathrm{evap}}$ of 60/90 °C, 90/60 °C and 90/90 °C.

The experimental investigations with the two storage units have proven the feasibility of adsorption TES for energy efficient heat supply of the brewing process [107]. We use these experiments as a basis for our study in the following sections.

4.2 Storage-unit modeling and adsorber calibration

In this section, we present the dynamic model of the adsorption TES unit with the evaporator and the adsorber described in Chapter 3. The general model equations are described in Section 4.2.1. The assumptions of the model are given in Section 4.2.2. The adsorber model is calibrated to measurement data (Section 4.2.3).

4.2.1 Mathematical model of the adsorption thermal energy storage unit

The dynamic model of the adsorption TES unit consists of an adsorber model (A) coupled to an evaporator model (E) (cf. Figure 4.3). Both models of adsorber and evaporator are coupled to external enthalpy flows via heat exchangers. Further heat and mass-transfer

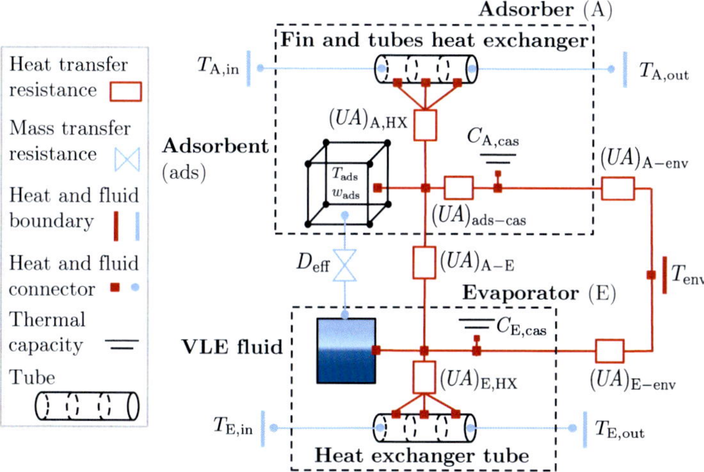

Figure 4.3: Scheme of the adsorption TES model with adsorber (A) and evaporator (E). The adsorbent (ads) is connected to the mass-transfer resistance with the effective coefficient D_{eff} and to three heat-transfer resistances with the following coefficients: $(UA)_{A,HX}$ connects to the adsorber heat exchanger, $(UA)_{ads-cas}$ connects to the casing heat capacity $C_{A,cas}$, and $(UA)_{A-E}$ connects to the evaporator (E). The evaporator casing $C_{E,cas}$ and the VLE fluid are connected to the environment (env) via the heat transfer with the coefficient $(UA)_{E-env}$ and to the evaporator heat exchanger via the heat transfer with the coefficient $(UA)_{E,HX}$. The VLE fluid is further connected to the adsorber via the mass-transfer resistance (D_{eff}).

resistances[1] connect adsorber, evaporator and the environment. The capacity-resistance model is written in the object-oriented modeling language Modelica [116], by using the TIL Suite [117] and the Adsorption Energy Systems Library [118] developed at the Institute of Technical Thermodynamics of RWTH Aachen University.

Adsorber model

The adsorber model contains the adsorbent model coupled to heat and mass transfer resistances and to the adsorber heat exchanger (cf. Figure 4.3). The adsorbent model includes the adsorbent material (ads), which is zeolite 13 X and water in adsorbed state

[1]The resistance is the reciprocal of the conductance [115]. In this thesis, we use the heat-transfer coefficients (UA) to describe the thermal conductance. Mass transfer is described with the effective coefficient D, which is also reciprocal to the resistance.

(ad). The adsorbent is described by a lumped model, i.e. the zeolite and the water in adsorbed state have the same temperature $T_{\mathrm{ad}} = T_{\mathrm{ads}}$ and pressure p_{ad} without spatial gradients. The density ϱ_{ad} and the heat capacity c_{ad} of the water in adsorbed state are assumed to be equal to that of liquid water.

The adsorbent/adsorbate equilibrium data in our library [118] is described as function of temperature T, pressure p and water loading $w = \frac{m_{\mathrm{ad}}}{m_{\mathrm{ads}}}$. The equilibrium data for zeolite 13 X and water is described with a characteristic function according to the Dubinin theory [119].

The mass-transfer resistance between the water vapor phase and the water in adsorbed state (ad) is described with an effective mass-transfer coefficient D_{eff}. Without a valve between the adsorber and evaporator (Section 4.1), the pressure inside the adsorber casing (p_{A}) is equal to the pressure in the evaporator (p_{E}). This pressure is defined by the temperature of the adsorptive water in vapor-liquid equilibrium (VLE) in the evaporator:

$$p_{\mathrm{A}} = p_{\mathrm{E}} = p_{\mathrm{sat}}(T_{\mathrm{VLE}}). \tag{4.2}$$

In the following, the equations of the mass and energy balances of the adsorption TES unit are shown. The signs in the equations are given for the adsorption phase with vaporization of the water. The same equations are valid for the desorption phase with condensation of the water, because the temperature and pressure differences change signs together with the heat and mass flows changing direction. The mass balance of the adsorbent is given by

$$\frac{\mathrm{d}m_{\mathrm{ad}}}{\mathrm{d}t} = m_{\mathrm{ads}}\frac{\mathrm{d}w}{\mathrm{d}t} = \dot{m}_{\mathrm{E-A}} = \dot{m}_{\mathrm{ad}}, \tag{4.3}$$

where m_{ads} is the mass of the adsorbent material and m_{ad} is the mass of water in adsorbed state. $\dot{m}_{\mathrm{E-A}}$ is the mass flow rate of adsorptive water from the evaporator to the adsorber. The adsorptive water is assumed to be completely adsorbed by the zeolite, because the mass in the vapor volume of the adsorber is negligible. The mass flow rate of adsorptive water $\dot{m}_{\mathrm{ad}} = \dot{m}_{\mathrm{E-A}}$ is described according to the linear driving force approach by Glueckauf [120] for spherical beads

$$\dot{m}_{\mathrm{ad}} = \frac{15 m_{\mathrm{ads}}}{{r_{\mathrm{ads}}}^2} D_{\mathrm{eff}} \left(w(T_{\mathrm{ads}}, p_{\mathrm{ad}}) - w(T_{\mathrm{ads}}, p_{\mathrm{E}})\right) \tag{4.4}$$

with the radius of the adsorbent beads r_{ads} and an effective mass-transfer coefficient D_{eff}. The driving force for the mass flow is described by the difference between the actual water loading of the adsorbent $w(T_{\mathrm{ads}}, p_{\mathrm{ad}})$ and the equilibrium loading $w(T_{\mathrm{ads}}, p_{\mathrm{E}})$ at the pressure of the vapor phase p_{E}.

The energy balance of the adsorber (A) reads

$$\begin{aligned}\frac{\mathrm{d}U_\mathrm{A}}{\mathrm{d}t} &= m_\mathrm{ads}\frac{\mathrm{d}}{\mathrm{d}t}\left[\left(c_\mathrm{ads}+wc_\mathrm{ad}\right)T_\mathrm{ads}\right] + m_\mathrm{A,lam}c_\mathrm{steel}\frac{\mathrm{d}T_\mathrm{ads}}{\mathrm{d}t}\\ &= \dot{m}_\mathrm{ad}h^v_\mathrm{VLE} - \dot{Q}_\mathrm{A,HX} - \dot{Q}_\mathrm{ads-cas} - \dot{Q}_\mathrm{A-E}\end{aligned} \tag{4.5}$$

with the specific internal energy of water in the adsorbed state u_ad and the enthalpy of the evaporator fluid in vapor phase h^v_VLE. The change of internal energy of the steel lamellae (A,lam) in the adsorber is also taken into account with the heat capacity of steel c_steel.

The heat flow rates are calculated by

$$\dot{Q} = UA\,\Delta T \tag{4.6}$$

with heat-transfer coefficients UA and the driving temperature difference ΔT.

The heat flow rate between adsorbent and the adsorber heat exchanger tubes $\dot{Q}_\mathrm{A,HX}$ is described with a constant effective heat-transfer coefficient $(UA)_\mathrm{A,HX}$:

$$\dot{Q}_\mathrm{A,HX} = (UA)_\mathrm{A,HX}\frac{1}{n}\sum_{i=1}^{n}\left(T_\mathrm{ads} - T_\mathrm{A,HX,i}\right), \tag{4.7}$$

with the temperature of the heat exchanger tube wall $T_\mathrm{A,HX}$.

The heat-transfer coefficient $(UA)_\mathrm{A,HX}$ and the mass-transfer coefficient D_eff are effective coefficients, because they combine different transfer mechanisms: the mass transfer combines mass transfer in the boundary layer and inside the pores of the zeolite and is thus difficult to predict [93]. A discussion of the different mass-transfer mechanisms during adsorption in porous solids can be found in [121]. The heat transfer combines contact heat transfer with conduction inside the zeolite and convective terms. Predicting the heat-transfer coefficient is also difficult, due to the complex geometry of the adsorber heat exchanger with unknown contact area $A_\mathrm{A,HX}$ between adsorbent and metal. To avoid such complicated and uncertain models, both the heat-transfer coefficient $(UA)_\mathrm{A,HX}$ and the mass-transfer coefficient D_eff are calibrated by fitting a simulation to measurement data.

The adsorber heat exchanger (Figure 4.3, top) is discretized and modeled as a tube from the TIL library [117]. This model accounts for the masses of the heat exchanger tubes and calculates a tube-side heat transfer according to Nusselt correlations [122].

The heat flow rate to the adsorber $\dot{Q}_{\text{des/ads}}$ during charging (des) and discharging (ads), which is used to evaluate the storage performance, is calculated from an energy balance around the adsorber heat exchanger, leading to

$$\dot{Q}_{\text{des/ads}} = \dot{m}_{\text{A,HX,oil}} \left(h_{\text{A,in}} - h_{\text{A,out}} \right) - \frac{\mathrm{d}U_{\text{A,HX,oil}}}{\mathrm{d}t} \tag{4.8}$$

with the enthalpy difference of the heat-transfer oil $\dot{m}_{\text{A,HX,oil}} \left(h_{\text{A,in}} - h_{\text{A,out}} \right)$ and the change of internal energy of the oil $U_{\text{A,HX,oil}}$ inside the adsorber heat exchanger.

Evaporator model

The evaporator model (Figure 4.3, bottom) is based on a vapor-liquid-equilibrium (VLE) model for the adsorptive water (E,ad) from the TIL Suite [117]. The evaporator volume $V_{\text{E}} = V_{\text{E,ad}}$ is filled by the VLE fluid with a total mass of $m_{\text{E,ad}}$. The mass balance is described by

$$\frac{\mathrm{d}m_{\text{E,ad}}}{\mathrm{d}t} = V_{\text{E}} \frac{\mathrm{d}\varrho_{\text{E,ad}}}{\mathrm{d}t} = -\dot{m}_{\text{ad}}. \tag{4.9}$$

We assume a homogeneous temperature $T_{\text{E,ad}} = T_{\text{E}}$ of the evaporator fluid and the evaporator metal parts (steel trays and casing). Accordingly, the internal energy of those metal parts (E,met) is included in the evaporator energy balance

$$\frac{\mathrm{d}U_{\text{E,ad}}}{\mathrm{d}t} + m_{\text{E,met}} c_{\text{steel}} \frac{\mathrm{d}T_{\text{E}}}{\mathrm{d}t} = -\dot{m}_{\text{ad}} h^{v}_{\text{E,ad}} + \dot{Q}_{\text{E,HX}} - \dot{Q}_{\text{E-env}} - \dot{Q}_{\text{A-E}} \tag{4.10}$$

with the internal energy of the adsorptive water $U_{\text{E,ad}}$. We also consider a heat loss to the environment $\dot{Q}_{\text{E-env}}$ which is calculated according to Equation (4.6).

The evaporator heat exchanger tube (Figure 4.3, bottom) is discretized and modeled as a tube from the TIL library [117]. The tube masses are included in the evaporator heat exchanger model which also calculates the tube-side heat transfer according to Nusselt correlations [122].

The heat flow rate between the adsorptive water and the heat exchanger surface in the evaporator

$$\dot{Q}_{\text{E,HX}} = (UA)_{\text{E,HX}} \frac{1}{n} \sum_{i=1}^{n} \left(T_{\text{E,HX,i}} - T_{\text{E}} \right) \tag{4.11}$$

is described using the heat-transfer coefficient $(UA)_{\text{E,HX}}$.

The heat flow rate to the evaporator $\dot{Q}_{\mathrm{cond/evap}}$ during charging (cond) and discharging (evap), which is used to evaluate the storage performance, is calculated from an energy balance around the evaporator heat exchanger, leading to

$$\dot{Q}_{\mathrm{cond/evap}} = \dot{m}_{\mathrm{E,HX,w}} \left(h_{\mathrm{E,in}} - h_{\mathrm{E,out}}\right) - \frac{\mathrm{d}U_{\mathrm{E,HX,w}}}{\mathrm{d}t} \tag{4.12}$$

with the enthalpy difference $\dot{m}_{\mathrm{E,HX,w}} \left(h_{\mathrm{E,in}} - h_{\mathrm{E,out}}\right)$ and the change of internal energy of the water $U_{\mathrm{E,HX,w}}$ inside the evaporator heat exchanger.

Heat losses

The heat losses of the storage unit are described with the following heat flows (cf. heat-transfer resistances in Figure 4.3): starting from the adsorber, heat flows to the casing ($\dot{Q}_{\mathrm{ads-cas}}$) and further to the environment ($\dot{Q}_{\mathrm{A-env}}$). In addition, for the storage unit with adsorber and evaporator in one container (cf. Figure 3.3, a), we have to consider a heat flow from the adsorber to the evaporator with $\dot{Q}_{\mathrm{A-E}}$. The heat loss of the evaporator is $\dot{Q}_{\mathrm{E-env}}$. All these heat-loss flow rates are calculated according to Equation (4.6).

4.2.2 Model assumptions

The heat-transfer coefficients for the heat-loss flow rates cannot be determined from transient measurements (cf. Section 2.4.2). However, steady-state measurements are not possible with the experimental setup in Figure 4.2, due to the low resolution of low heat flow rates (cf. Appendix B.1) that occur during steady state. We hence calculate the heat-loss coefficients according to Nusselt correlations [122] as specified in Table 4.1. To calculate the heat-transfer coefficients from the adsorber $(UA)_{\mathrm{A-env}}$ and evaporator $(UA)_{\mathrm{E-env}}$ to the environment, we assume average temperatures of the adsorber casing and evaporator casing. For the convective heat transfer inside the adsorber casing (cf. Table 4.1), we assume an average velocity of the vapor flow and an average temperature of the adsorber surface. As a result, the heat-transfer coefficients are constant.

The calibration procedure for the heat-transfer coefficient $(UA)_{\mathrm{A,HX}}$ and the mass-transfer coefficient D_{eff} in this chapter is explained in Section 4.2.3. The results are listed in Table 4.1.

The heat transfer in the evaporator is assumed to be ideal, which we define by a sufficiently high coefficient $(UA)_{\mathrm{E,HX}}$ for vaporization and condensation to ensure that the process

Table 4.1: Transfer coefficients of the adsorption TES model to describe the performance of two adsorption TES units

Coefficient	Value	obtained by
D_{eff}	$1 \times 10^{-10}\,\mathrm{m^2/s}$	calibration to experimental data (cf. Section 4.2.3)
$(UA)_{\mathrm{A,HX}}$	$480\,\mathrm{W/K}$	calibration to experimental data (cf. Section 4.2.3)
$(UA)_{\mathrm{ads-cas}}$	$26.16\,\mathrm{W/K}$	radiative heat transfer according to [123] eq. 8-37 and convective heat transfer according to [122] FA2, eq. 12
$(UA)_{\mathrm{A-E}}$	$0.748\,\mathrm{W/K}$	radiative heat transfer according to [123] eq. 8-37 and convective heat transfer according to [122] Fa4, eq. 21
$(UA)_{\mathrm{A-env}}$	$1.688\,\mathrm{W/K}$	according to [122] Eb1, eq. 1
$(UA)_{\mathrm{E-env}}$	$0.914\,\mathrm{W/K}$	according to [122] Eb1, eq. 1
$(UA)_{\mathrm{evap}}$	$1400\,\mathrm{W/K}$	not limiting
$(UA)_{\mathrm{cond}}$	$2000\,\mathrm{W/K}$	not limiting

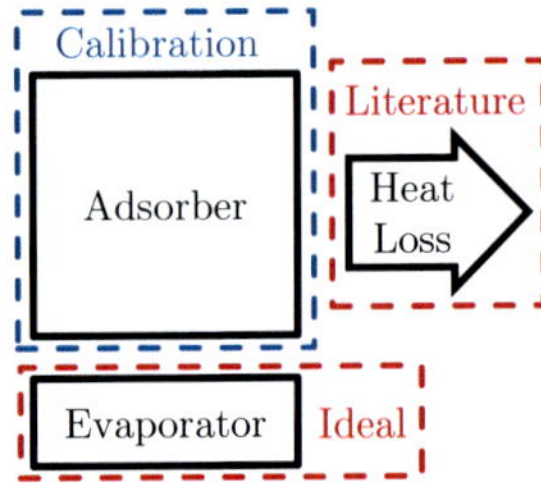

Figure 4.4: Identification of model parameters in Chapter 4: adsorber heat and mass-transfer coefficients are experimentally calibrated, evaporator is assumed to be ideal without heat and mass transfer limitations, heat-loss coefficients are derived from literature.

on the tube surface is not limiting the heat transfer. This was done since the evaporator heat flow rates (Equation (4.1)) could not properly be resolved, because the temperature differences across the evaporator heat exchanger were below the measurement uncertainty (cf. Appendix B.1).

Neglecting the limitations of evaporator and condenser is in accordance with, e.g. Fernandes et al. [94]. Accordingly, the temperature of the evaporator heat exchanger dictates the adsorptive water vapor pressure in the system. As a result, the adsorber is the only component to limit the adsorption process by heat and mass-transfer resistances.

Figure 4.4 provides an overview of the means for parameter identification for the model used to simulate the brewing process. Table 4.1 summarizes the results for all heat and mass-transfer coefficients of the adsorption TES model used in this chapter. It has to be noted that the values of the heat-transfer coefficients scale with the size of the storage system, i.e. with the number of units: since the model has been calibrated to measurements with two storage units (cf. Figure 4.2), the heat-transfer coefficients also relate to two adsorbers and two evaporators each.

In this chapter, we describe the equilibrium of zeolite 13 X and water with the characteristic function, fitted to experimental data by Núñez [110]. In accordance with Núñez, the heat capacity of the zeolite is assumed to be constant during simulations used in this chapter.

4.2.3 Adsorber calibration

The obtained experimental data (cf. Section 4.1) is used to calibrate the model of the adsorption TES unit, in particular the adsorber model as the critical component for storing and releasing heat. The heat and mass-transfer coefficients in the adsorber model are calibrated by comparing the experimental and simulated heat flow rates of the adsorber.

Without measurement of conditions inside the storage unit, heat and mass transfer parameters of adsorption systems cannot be identified separately [124]. Thus, the parameters are determined simultaneously by fitting simulation and measurement data in accordance with Lanzerath [93]. For this purpose, we minimize the coefficient of variation (CV, cf. Equation (2.5)) based on the root mean square deviation ($RMSD$) between the measured heat flow rate $\dot{Q}_{\text{meas}}$ and the simulated heat flow rate $\dot{Q}_{\text{sim}}$ in the adsorber (cf. Equation (4.8)). The fitting was carried out for the measurements described in Section 4.1: in particular individually for the three temperature combinations $T_{\text{cond}}/T_{\text{evap}}$ in the evaporator unit: 60/90 °C, 90/60 °C and 90/90 °C. All three fitted parameter sets were identical within measurement tolerance. Thus, the coefficients for heat and mass transfer in the adsorber are assumed constant over the regarded temperature range.

Figure 4.5 shows an example of measured and simulated heat flow rates for both charging and discharging of the adsorption TES unit. Measured and simulated heat flow rates in the adsorber correspond very well, except the simulation slightly underestimates the discharging heat flow rate during the last ten minutes of adsorption (cf. Figure 4.5). This underestimation of the adsorber heat flow rate is most likely due to an overestimation of the heat losses from the adsorber to the casing and to the environment. The heat losses tend to be overestimated at the end of adsorption, because they are calculated with constant heat-transfer coefficients that were determined for a constant temperature of the adsorber surface and the adsorber casing (cf. Section 4.2.2). The real temperatures

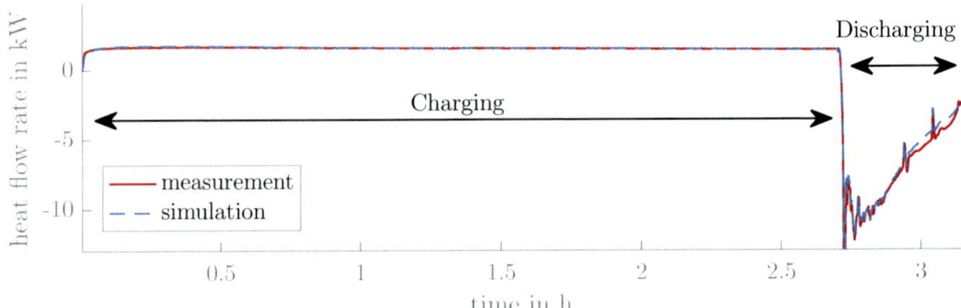

Figure 4.5: Comparison of measured and simulated heat flow rates $\dot{Q}_{\mathrm{A,HX}}$ (cf. Equation (4.7)), of the adsorber during charging and discharging. Example measurement with condensation at 60 °C, desorption until 200 °C and adsorption at 120 °C, vaporization at 90 °C.

are lower at the end of adsorption than the average value used for the calculation of the heat-transfer coefficient.

In all measurements, the coefficient of variation CV of the adsorber heat flow rate is around 10 %, which is within the measurement uncertainty (cf. Appendix B.1). Accordingly, the calibrated model is regarded sufficient to describe the adsorption TES behavior at conditions similar to the experimental investigation. Based on our experience with adsorption-based energy systems [82], we expect extrapolation to give reasonable results: we expect the simulations to provide at least a qualitatively good description of the behavior of the adsorption TES unit, even though the model is not validated for evaporator temperatures below 60 °C, i.e. for vapor pressures below 200 mbar.

4.3 Storage integration study

In this section, we discuss the integration of adsorption thermal energy storage (adsorption TES) into a brewing process. The process setup is introduced in Section 4.3.1. We define measures for the energy recovery ratio to evaluate the performance of our adsorption TES unit in Section 4.3.2.

4.3.1 Process setup

We investigate the performance of adsorption TES combined with a gas-powered cogeneration unit for batch-process energy supply. The cogeneration unit of this study is a biogas-driven engine, type E2842 by MAN with a power output of 250 kW, a mechanical efficiency of 37.5 % and a thermal efficiency of 54.5 % at full load [125]. The process-heat demand structure is taken from the batch brewing process as shown in Figure 4.1 with a process-heat demand of 380 kWh. This process-heat demand is met by the combined heat output of the adsorption TES unit during discharging and the cogeneration exhaust gas at the same time (cf. Figure 4.6). To match the heat demand for the given brewing process, the simulation assumes 135–210 units of our adsorption TES system to be charged and discharged in parallel. In the following, these adsorption TES units are referred to as one single unit.

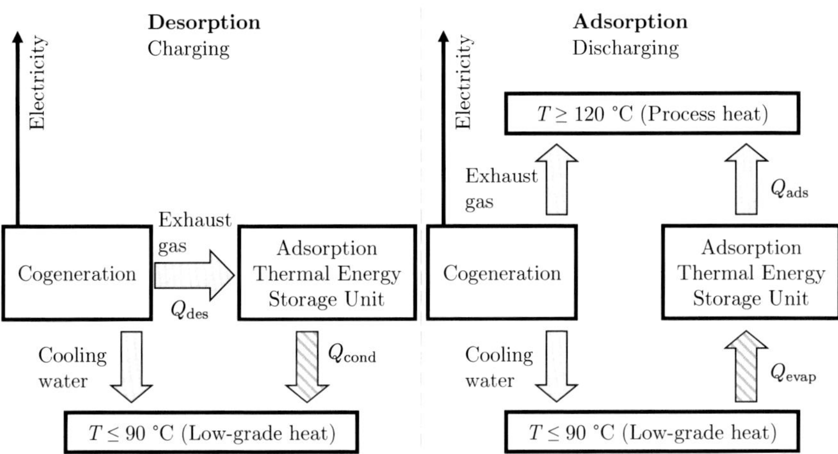

Figure 4.6: Heat flow chart of process-heat supply by cogeneration and adsorption thermal energy storage during desorption (left) and adsorption (right).

The storage process is as follows: during charging, the adsorption TES unit receives the heat Q_{des} from the cogeneration exhaust-gas for desorption (Figure 4.6 left). At the same time, the heat Q_{cond} from condensation is released to the environment or can be used by the process at a temperature below 90 °C. During adsorption, the released heat Q_{ads} is used for process-heat supply at a temperature of 120 °C (Figure 4.6 right). At the same time, the heat Q_{evap} has to be provided to the storage unit for vaporization. Since the adsorption

TES unit uses only the high temperature exhaust-gas from the cogeneration unit, low temperature cooling water from the cogeneration unit is still available. We therefore assume that the heat from the cooling water is used in the process at temperatures below 90 °C (cf. Figure 4.6).

4.3.2 Performance of the adsorption thermal energy storage unit

The suggested energy supply system potentially leads to primary energy savings due to the efficiency of cogeneration. The achievable energy savings are dependent on the storage energy recovery ratio ERR, as defined in Equation (2.3). The energy recovery ratio of adsorption TES strongly depends on the heat structure of the process, i.e. temperatures, time profiles and existence of heat flows. In this storage integration study, we focus on the structure of the low-grade heat, in particular whether a demand exists for low temperature heat from condensation (C±) and whether excess heat is freely available for vaporization (E±). The resulting options to integrate low-grade heat inputs and outputs of adsorption TES are reflected in the calculation of the energy recovery ratio. For our study in this chapter, we define four cases of the energy recovery ratio, which we call $\eta_{\mathrm{E+C+}}$, $\eta_{\mathrm{E+C-}}$, $\eta_{\mathrm{E-C+}}$ and $\eta_{\mathrm{E-C-}}$ as defined in Table 4.2.

Table 4.2: Energy recovery ratio $\eta = \frac{Q_{\mathrm{used}}}{Q_{\mathrm{effort}}}$ of adsorption thermal energy storage depending on low-grade heat supply and demand; calculation distinguishes whether a demand exists for low temperature heat from condensation (C+) or not (C-) and whether excess heat for vaporization is freely available (E+) or not (E-)

		Condenser	
		Heat is used (C+)	Heat is not used (C-)
Evaporator	Heat is freely available (E+)	$\eta_{\mathrm{E+C+}} = \frac{Q_{\mathrm{ads}}+Q_{\mathrm{cond}}}{Q_{\mathrm{des}}}$	$\eta_{\mathrm{E+C-}} = \frac{Q_{\mathrm{ads}}}{Q_{\mathrm{des}}}$
	Heat has to be provided (E-)	$\eta_{\mathrm{E-C+}} = \frac{Q_{\mathrm{ads}}+Q_{\mathrm{cond}}}{Q_{\mathrm{des}}+Q_{\mathrm{evap}}}$	$\eta_{\mathrm{E-C-}} = \frac{Q_{\mathrm{ads}}}{Q_{\mathrm{des}}+Q_{\mathrm{evap}}}$

The four energy recovery ratios η of the adsorption TES unit are determined for the given industrial batch process using the calibrated model presented in Section 4.2. The heats Q are calculated by integrating the respective heat flow rates $\dot{Q}$ from the dynamic simulation (cf. Equations (4.8) and (4.12)).

In the storage integration study, we fix the structure of the process-heat demand from the brewing process and the structure of charging by the cogeneration unit according to Figure 4.1. The low-grade heat structure is varied to examine various possible demand and supply structures. This variation of heat structure allows investigating the influence

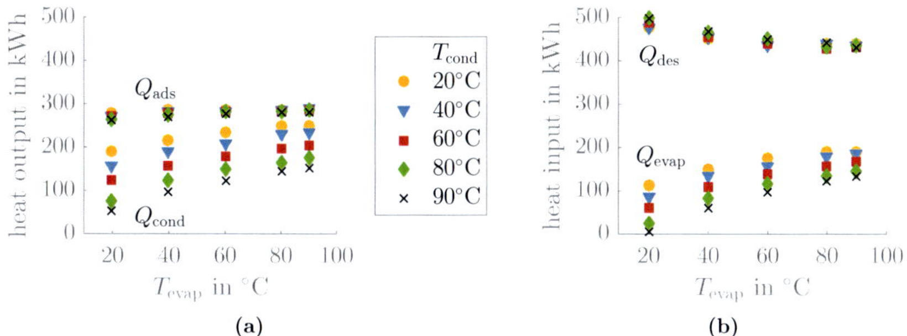

Figure 4.7: Heat output (a) and input (b) of the adsorption thermal energy storage unit depending on vaporization temperature T_{evap} and condensation temperature T_{cond}.

of variable conditions for vaporization and condensation on the energy recovery ratio of the adsorption TES unit.

In particular, we vary the temperatures of vaporization T_{evap} and condensation T_{cond} from 20 °C–90 °C for the storage integration study. Figure 4.7 shows the resulting heat outputs Q_{ads} and Q_{cond} and inputs Q_{des} and Q_{evap} of the adsorption TES unit. The heat output at process temperature during adsorption Q_{ads} is nearly constant as the process-heat demand is fixed to 380 kWh (cf. Section 4.3.1). This heat demand is met by the adsorption TES unit and the cogeneration exhaust gas with a ratio of

$$\frac{Q_{ads}}{Q_{cogeneration}} = 69 - 76\,\%, \tag{4.13}$$

which varies slightly with the low-grade heat temperature.

While the discharging heat varies only slightly, charging the storage unit requires more heat Q_{des} for a lower vaporization temperature T_{evap}. The dependence on the low temperature is even more visible for Q_{evap} and Q_{cond}: the higher T_{evap} and the lower T_{cond}, more heat Q_{evap} is needed during vaporization (cf. Figure 4.7 (b)), but at the same time more heat is released during condensation Q_{cond} (cf. Figure 4.7 (a)).

Based on the heat in- and outputs (cf. Figure 4.7), the energy recovery ratio of the adsorption TES unit was determined for all cases defined in Table 4.2. As expected, the results strongly depend on the low-grade heat demand and supply structure of the process. In addition, the temperatures of the low-grade heat influence the energy recovery ratios $\eta_{E\pm C\pm}$. Thus, the results are shown as a function of the temperature of vaporization and condensation in Figure 4.8. η_{E+C-} and η_{E-C-} are not evaluated for variable condensation temperatures, since the heat from condensation is not used in these cases. It is therefore not necessary to condense at elevated temperatures.

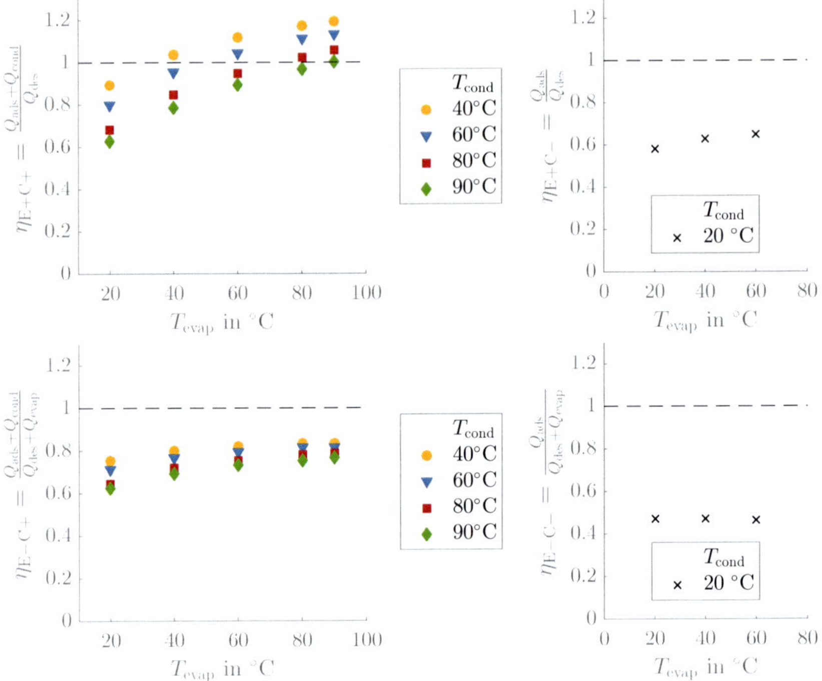

Figure 4.8: Storage energy recovery ratios $\eta_{E\pm C\pm}$ for different heat integration options (cf. Table 4.2) depending on vaporization temperature T_{evap} and condensation temperature T_{cond}.

The recovery ratios η_{E+C+} and η_{E-C+} are always higher than η_{E+C-} and η_{E-C-}, because heat from condensation is used. The energy recovery ratio η_{E+C+} even reaches values greater than 1 due to the use of condensation enthalpy combined with heat freely available for vaporization.

For both E+ cases, i.e. when heat for vaporization is for free, energy recovery increases with vaporization temperature T_{evap}, since a higher temperature leads to a higher pressure in the storage unit. Following the adsorption equilibrium, more water is adsorbed leading to a proportionately higher process heat output Q_{ads}. At the same time, more water has to be evaporated, leading to an increasing Q_{evap}. Additionally, heating-up the evaporator has to be taken into account at the beginning of each adsorption phase, if the vaporization temperature is higher than the condensation temperature. This additional heat also leads to an increasing Q_{evap}. Thus, for both E- cases, the increasing Q_{evap} counterbalances an increase of Q_{ads} for rising T_{evap}.

A decrease in condensation temperature T_{cond} increases the energy recovery ratio: with a lower pressure during charging, more water is desorbed. Thus, more heat can be stored in the adsorber. At the same time, heat losses from the adsorber to the condenser will rise with falling condensation temperature, as the temperature difference increases. The heat losses between adsorber and condenser are a significant disadvantage of the storage unit using only one container without valve (cf. Figure 3.3).

The study shows that adsorption TES is most beneficial if low grade excess heat is available at times of discharging and if low-grade heat is required by the process when charging the storage unit (case E+C+ in Figure 4.8). In this case, adsorption TES benefits from the heat pump effect. On the contrary, in case E-C-, the heat from condensation is released to the environment and extra heat has to be provided for vaporization: here, the maximum achievable storage energy recovery ratio, neglecting heat losses, is in the range of 55 %. This value of the energy recovery ratio corresponds well to the enthalpy of vaporization/condensation being approximately 5/6 of the enthalpy of adsorption for zeolite 13 X and water

$$\eta_{\text{E-C-}} = \frac{Q_{\text{ads}}}{Q_{\text{des}} + Q_{\text{evap}}} \approx \frac{1}{1 + 5/6} = 0.55. \tag{4.14}$$

In case E-C-, adsorption TES leads to large energy losses as it is not possible to use the low-grade heat output Q_{cond}. Nevertheless, for long-term TES, even the case E-C- can be reasonable as the adsorption enthalpy is not prone to sensible heat losses and thus offers a unique feature that other storage technologies cannot offer.

In summary, adsorption TES can achieve high energy recovery ratios, even reaching values above one. However, the achievable energy recovery ratio has to be evaluated individually for every case.

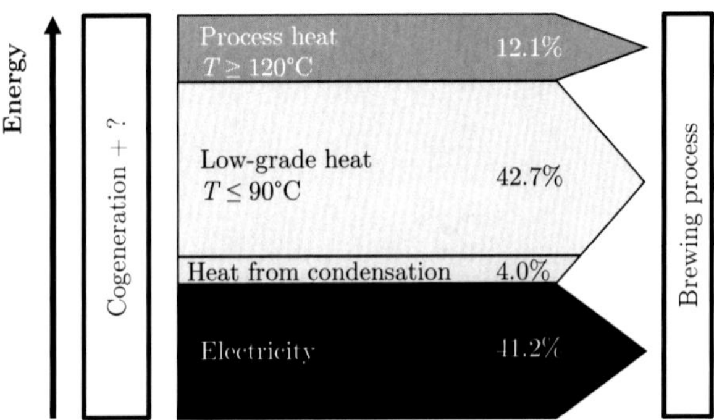

Figure 4.9: Qualitative energy-balance diagram for cogeneration-integration options. Energy-supply arrows drawn to scale. In case of cogeneration integration with adsorption thermal energy storage, a small part of the low-grade heat can be provided by condensation. Numeric percentages are valid for the temperature combination 90/60 °C for condensation/vaporization.

4.4 Primary energy savings and comparison of alternative integration concepts

Cogeneration leads to savings in primary energy due to the high efficiency of simultaneously producing electrical power and heat. These savings are, however, only achieved if there is also a demand for both heat and power. Thus, if integrated into batch processes, cogeneration is preferably integrated together with a secondary technology balancing the time-varying heat demand. Adsorption thermal energy storage (TES) has been shown to successfully shift heat in time and thus to satisfy the demand of a process (Section 4.3). Still, alternative technologies are available to compensate variations in heat demand. In this chapter, we compare adsorption TES to both peak boilers and latent TES. Each alternative is assumed to support a cogeneration unit for process energy supply. The study is based on the best case for the adsorption TES energy recovery ratio η_{E+C+} in Table 4.2, i.e. excess heat is available for vaporization and heat from condensation is used in the process. For comparison, all three cogeneration integration options have to deliver the same amount of energy; specifically process heat, low-grade heat and electricity (Figure 4.9).

The performance of the peak boiler and the latent TES unit are modeled with steady-state models using literature data. The boiler efficiency is set to 90 % [126]. The latent TES system uses high-density polyethylene with a melting temperature of 128 °C as storage material. The energy recovery ratio of this latent TES unit is assumed to be 86 % [127]. No further dynamics of the latent TES unit are taken into account. Moreover, it is assumed that any desired heat flow rate is always deliverable by the latent TES unit; therefore, no restrictions concerning heat transfer in the phase change material are considered.

Additional electricity generation is taken into account by grid energy supply from the German power mix with a primary energy efficiency of 42 % [128].

For evaluation of adsorption TES, we use the calibrated adsorption TES model (cf. Section 4.2.3). Consequently, dynamics are taken into account with all inherent restrictions: the heat flow rate from the storage unit is limited due to the low heat-transfer coefficient of the storage material. Heat losses reducing the energy recovery ratio are also included and estimated to be significant for the given time profile with the long duration of charging. For the adsorption TES unit, a thermal insulation is considered based on a standard mineral wool with 5 mm thickness and a thermal conductivity of 0.06 W/(m K) [129].

The process-heat demand structure is taken from the brewing process (Figure 4.1). In addition, the low-grade heat demand is assumed to be approximately three times the process-heat demand (cf. Figure 4.9). Furthermore, the heat from condensation provided by the adsorption TES unit is supposed to be used in the process. However, this heat corresponds to only 3 to 20 % of the total low-grade heat from cogeneration (Figure 4.9).

The variable heat flow rate of the condenser is due to the performance of the adsorption TES unit which varies with the low-grade heat temperature (cf. Section 4.3.2). Both the size of the adsorption TES unit and the size of the cogeneration unit are adapted, because the process-heat supply is fixed. The electrical power output of the cogeneration unit thus changes in the range of 200–240 kW depending on the low-grade heat temperatures.

To evaluate energy supply efficiency, the primary energy consumption is calculated for three options: (1) the combination of cogeneration and adsorption TES is used as reference and compared (2) to cogeneration plus peak-load boiler and (3) to cogeneration plus latent TES. The results of the comparison are illustrated by the ratio of primary energy consumption of the regarded alternative (2 and 3) to primary energy consumption of the reference combination (1) (Figure 4.10). Thus, a ratio above 100 % implies that cogeneration with adsorption TES is more efficient in primary energy consumption compared to the alternative.

For adsorption TES, the primary energy efficiency inherits the dependence on the temperature of condensation and vaporization, as these temperatures determine the energy recovery ratio of the storage unit (see Section 4.3): the system with adsorption

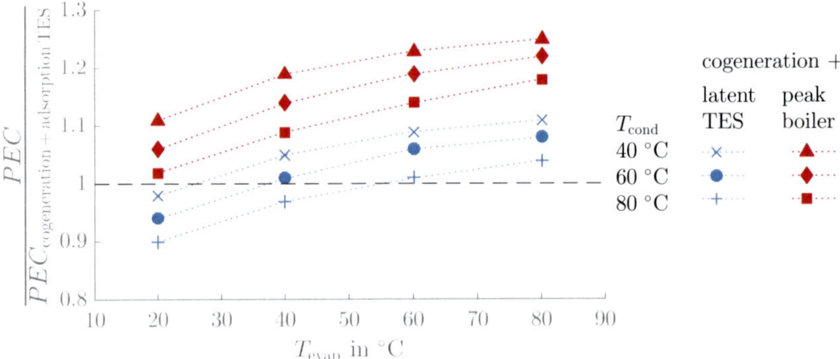

Figure 4.10: Primary energy consumption (PEC) for cogeneration with peak boiler and with latent thermal energy storage (TES) compared to cogeneration with adsorption TES. The ratio is plotted against temperatures of vaporization T_{evap} for condensation temperatures T_{cond} varied between 40 and 80 °C. A ratio above 100 % implies that adsorption TES is more energy efficient than the alternative (lines just guide the eye).

TES achieves high efficiencies for vaporization at elevated temperature and condensation at low temperature. For the best case, cogeneration with adsorption TES can reduce primary energy consumption by up to 25 % compared to the combination of a cogeneration unit with a peak boiler. Figure 4.10 shows that for all cases the combination of cogeneration with TES technologies is energetically beneficial compared to a peak boiler. Cogeneration can provide more heat and thus more electricity when combined with thermal energy storage. Consequently, a solution with thermal energy storage is superior to a peak boiler.

For the TES technologies considered, results depend on the structure of the low-grade heat in the process: latent TES improves the process energy efficiency by 10 % compared to adsorption TES for vaporization at 20 °C and condensation at 80 °C. However, with vaporization at 80 °C and condensation at 40 °C, adsorption TES outperforms latent TES. The process energy efficiency is improved by more than 10 %. Thus, the specific heat structure of a process has to be taken into account when deciding for the best TES option to integrate cogeneration in an industrial batch process.

4.5 Discussion of model assumptions

In the interpretation of this study, the underlying assumptions have to be considered for each technology. In particular, heat losses for latent thermal energy storage (TES) during the long charging time are underestimated, as the employed energy recovery ratio for latent TES was determined for short charging periods by Cottone and Mehling [127]. Furthermore, no heat flow rate restrictions were taken into account for the latent TES unit. Yet, phase change materials are known to suffer from poor heat-transfer characteristics [130]; accordingly, heat flow rate restrictions should be considered. It can thus be assumed that the presented results for latent TES are optimistic estimates.

Heat losses have been considered for the adsorption system. However, the heat-loss coefficients have been calculated based on Nusselt correlations and could not be experimentally validated (cf. Section 4.2.2). Yet, experimental validation is crucial to achieve a valid description of the heat losses, as we discussed in Section 2.4.2. A valid description of heat losses is even more important as the heat losses may significantly reduce the performance of the storage unit. As a consequence, the heat losses are further investigated in this thesis in Chapter 6.

4.6 Summary and motivation for the following chapters

The presented study evaluates the benefits of adsorption TES combined with cogeneration for energy supply of industrial batch processes. Based on previous experiments with an adsorption TES unit, a dynamic model is built to describe the behavior of the adsorption TES unit. The model is calibrated to measurement data to achieve reliable simulation results for the analysis on the system level.

A brewery is used as an exemplary industrial batch process with cogeneration energy supply. The process-heat demand of the brewery is satisfied by the cogeneration exhaust-gas and adsorption TES. We further study the influence of low-grade heat on the storage energy recovery ratio. The results of this chapter illustrate the strong dependence of adsorption TES performance on vaporization and condensation temperatures. Even more importantly, the availability of low-grade heat for vaporization is crucial. If low-grade heat is available, the energy recovery ratio can even rise above 100 %.

The primary energy consumption is estimated for three alternative concepts to integrate cogeneration into process energy supply: our adsorption based TES system is compared to a conventional system with a peak load boiler and an innovative system with latent TES. The study shows that thermal energy storage is always superior to a peak boiler: thermal energy storage can significantly reduce primary energy consumption for industrial batch-process energy supply. The choice of the best storage technology depends on the process-specific structure of heat supply and demand: integration of low-grade heat by adsorption TES can immensely enhance the process energy efficiency.

The discussion of the model assumptions (cf. Section 4.5) confirms that a comprehensive assessment of the benefits of TES for an energy system relies on an accurate description of the performance of the considered TES technology. The performance is determined by the dynamic properties of the TES system, in particular the achievable heat flow rates, and the heat losses.

Thus far, the heat losses cannot be determined by the measurements of the adsorption TES unit (cf. Section 4.2.2). Furthermore, the measurements of heat flow rates, and especially of the evaporator heat flow rates, are uncertain, because the temperature differences across the heat exchangers are below or close to the measurement uncertainty of the experimental setup for brewery investigations (cf. Section 4.1).

As a consequence, in the model, we assume an ideal evaporator and the heat losses are derived from literature (cf. Figure 4.4). However, all deviations that may be caused by model assumptions are adjusted during adsorber calibration: erroneous heat-loss coefficients and missing limitations in the evaporator model are compensated by the adsorber heat and mass-transfer coefficient. Although the model accurately describes the performance of the storage unit (cf. Figure 4.5), a thorough performance analysis remains difficult: the performance of the individual components, i.e. evaporator and adsorber, and the heat losses of the storage unit cannot be analyzed separately. However, optimization of a TES unit requires identification of bottlenecks in the performance of the components of a TES system.

Hence, the goal of the following chapters is to provide a model that accurately describes the performance of the adsorption TES unit while distinguishing between adsorber, evaporator and the interaction with the environment, i.e. heat losses. To accurately determine the heat flow rates and the heat losses of the adsorption TES unit, we build a new experimental setup (Chapter 5) with improved measurement accuracy and more measurement positions to enable thorough measurements of heat losses. With this new setup, a comprehensive experimental characterization of the performance of the adsorption TES unit is possible (Chapter 6). With the measurements of heat losses and storage cycle performance, we are able to thoroughly calibrate all relevant heat and mass-transfer coefficients in both the adsorber model and the evaporator model of our storage-unit model (Chapter 7).

5 A Versatile Experimental Setup for Thermal Energy Storage Units

The results from Chapter 4 emphasize that the assessment of an adsorption thermal energy storage (TES) system requires the precise knowledge of the TES performance. The performance is a result of the dynamic properties of the TES system, i.e. the achievable heat flow rates and the heat losses. The heat losses of a closed-adsorption-TES unit have not yet been accurately determined (cf. Section 2.4.2). Previous experimental investigations (cf. Section 4.1 and Reference [107]) have proven the feasibility of adsorption TES, but the measurement equipment dis not allow for the determination of heat losses (cf. Section 4.2.2): the temperature of the casing wall of the storage unit is not measured and the resolution of the existing measuring devices is not sufficient for low heat flow rates.

To allow for an accurate determination of heat losses and performance of the adsorption-TES unit from Chapter 3, we present a new experimental setup for storage characterization in Section 5.1. We further introduce the design of an adsorption-TES unit with a valve between adsorber and evaporator in Section 5.2.

5.1 Experimental setup

The performance characterization of adsorption TES requires accurate quantification of the heat flow rates in both parts of the storage unit, the adsorber and the evaporator. For a comprehensive experimental analysis, the relevant process conditions should be emulated, in particular the time delays for storage periods and the temperatures of charging and discharging.

Adsorption TES systems are applicable to various processes on different temperatures (cf. Sections 2.3and 2.5.5). However, a maximum temperature can be identified based on the equilibrium data of the storage working pair. The zeolite 13 X, which is applied in this thesis, is nearly dry at 250 °C [85]. As a result, for higher temperatures, heat is no longer stored by adsorption and the storage principle alters to purely sensible TES.

Thus, our zeolite-based setup is designed for storage measurements with temperatures up to 260 °C. To provide for the required working conditions, a thermal oil is used as the heat transfer fluid to allow for high temperatures without a pressurized system. We choose a thermal oil with a comparably low viscosity to ensure turbulent flow regimes and thus good heat transfer rates at most temperature conditions. More detail on the oil is provided in Appendix A.2.

Figure 5.1: Experimental setup with adsorption-TES unit (left) and oil and water circulation systems. Insulation is removed for better visibility of the single parts.

Figure 5.1 shows a picture of the experimental setup with the storage unit. The adsorber heat exchanger is connected to the circulation system of the thermal oil. For charging, the oil in the circulation system is heated by an electric heater with maximum output of 3 kW. During discharging, heat is released via a secondary heat exchanger to a water circulation system, allowing for temperatures up to 90 °C without a pressurized system.

The water circulation system is also equipped with an electric heater to provide heat to the evaporator. To cool the evaporator, the water circulation system is connected via a heat exchanger to the refrigeration cycle of the laboratory. Here, a minimum temperature of 9 °C can be provided.

The automatic control of all components in the circulation systems, including heaters, valves and variable-speed pumps, enables the user to emulate diverse storage cycles and steady-state conditions. Temperatures, heat flow rates and storage periods can be adjusted according to the application.

We measure the temperatures in the oil and water circulation systems by Pt100 resistance thermometers. The mass flow rate in the oil circulation system is measured with a Coriolis flow meter; volume flow in the water circulation system is measured with vortex flow meters. Manufacturer specifications of the measurement equipment can be found in Appendix A.

The chosen measurement equipment and the flow control allows for accurate measurements in the circulation systems connected to the heat exchangers. The measurement uncertainty is improved by a factor of 10–20 compared to the setup we used for calibration in Chapter 4. We are now able to determine heat flow rates with a maximum uncertainty of 45 W. More information on the determination of the measurement uncertainty is provided in Appendix B.

5.2 Thermal energy storage unit

With the presented experimental setup (Figure 5.1), we characterize the performance of the adsorption thermal energy storage (TES) unit. For the experimental and the model-based analysis in the following chapters, we vary the design of our adsorption-TES unit. In contrast to the storage unit studied in Chapter 4, we now separate the adsorber and evaporator with a butterfly valve (cf. Figures 3.3 (b) and 5.2). 5 cm of rock wool insulation

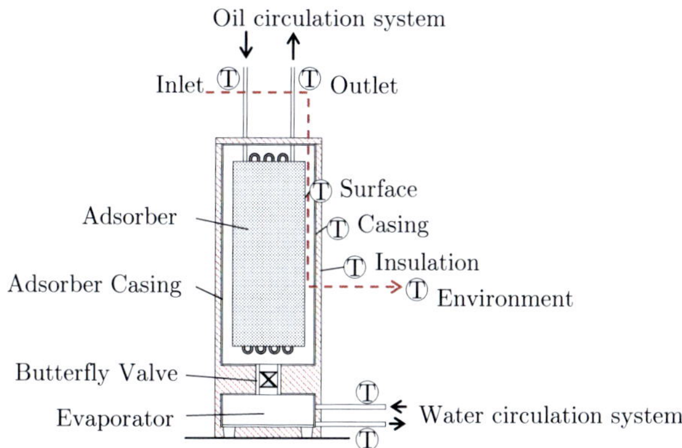

Figure 5.2: Scheme of adsorption thermal energy storage unit: adsorber and evaporator in separate vacuum-sealed casings and a valve in-between. Temperature measurement positions (T) – the dashed line shows the temperature measurement positions used to analyze heat losses of the adsorber in Chapter 6.

are applied to the outer surface of the adsorber casing and the evaporator. The total volume of the storage unit with insulation is $V_{\text{store}} = 0.148\,\text{m}^3$.

The butterfly valve between adsorber and evaporator ensures mechanical separation of water and zeolite during storage periods. As a result, the adsorption-TES unit can be applied to storage processes with storage periods between de- and adsorption. Additionally, the valve facilitates the measurement of equilibrium data: the adsorber can be separated from the evaporator and the weight can be measured to determine the equilibrium loading of the adsorber.

We measure the temperatures on the insulation, inside the evaporator and on the adsorber casing with Pt100 resistance thermometers. Thermocouples K-type are used to measure the ambient temperature and the temperature on the adsorber surface (cf. Figure 5.2). Additional Pt100 resistance thermometers and Thermocouples K-type are available for redundant measurement positions, e.g. on the surface of the adsorber casing, to control the temperature measurements and to check for inhomogeneous temperature distributions on the casing surface.

The measurement of the casing temperature and the improved measurement uncertainty allow for experimental characterization of heat losses in steady-state measurements. The experimental characterization of our adsorption-TES unit is described in the following Chapter 6.

6 Experimental Analysis of Storage Performance

This chapter provides a comprehensive analysis of the performance of our small-scale, closed adsorption thermal energy storage (TES) unit (cf. Section 5.2). Residential heating is considered as an application to define the discharging temperature. In contrast to the industrial application (Chapter 4), residential heating requires lower heat supply temperatures around 70 °C [131]. Even though the temperatures may be lower, storage periods between de- and adsorption increase the storage cycle time and lead to larger heat losses.

We use the following approach to systematically determine the heat losses of the adsorber and subsequently evaluate the storage performance: first, steady-state measurements at constant temperatures are analyzed to reveal the intrinsic heat losses of the storage unit (Section 6.1). We derive the heat-loss coefficients of the adsorber and compare the measured results to calculations based on Nusselt correlations. Moreover, we estimate the share of radiative losses in the total heat loss to test the hypothesis of Dawoud et al. [35] and Li et al. [33] who postulate that heat losses of a closed adsorption unit are mostly due to radiation.

Secondly, we examine the performance of our storage unit in cycle measurements for different charging temperatures (Section 6.2) to find the optimal storage temperature, as discussed in Section 2.4.1. We start with measurements where charging (desorption) of the storage unit is directly followed by discharging (adsorption). As the next step, we measure storage cycles with 2 hours of storage time between de- and adsorption. Based on the measured heat-loss coefficients, we calculate the heat losses during charging, discharging and storage to finally determine the thermal efficiencies of the storage unit. To provide a

---

Contents of this chapter have been published in:

H. Schreiber, F. Lanzerath, C. Reinert, C. Grüntgens, and A. Bardow. "Heat lost or stored: Experimental analysis of adsorption thermal energy storage". *Applied Thermal Engineering* 106 (2016), pp. 981–991.

lower bound for the energy recovery ratio, we also measure the extreme case of long-term storage.

To finally assess the results of our extensive storage-cycle measurements, we compare the energy storage density and energy recovery ratio in Section 6.3. Section 6.4 provides a summary to this chapter.

6.1 Heat-loss analysis from steady-state measurements

In this section, we deduce heat-loss coefficients from steady-state measurements and discuss the share of radiation in the total heat losses. Quantification of heat losses is possible with measurements at constant temperatures, since the storage unit is in steady-state condition and neither adsorption nor any other change of internal energy occurs. During steady state, the heat losses equal the thermal input to the storage unit:

$$\dot{Q}_{\text{steady-state, loss}} = \dot{Q}_{\text{steady-state, in}} \tag{6.1}$$

The thermal input to the storage unit during steady state is the heat transferred to the adsorber heat exchanger $\dot{Q}_{\text{steady-state, in}}$ which is equal to the change of enthalpy of the heat-transfer oil $\Delta\dot{H}_{\text{oil}}$ between adsorber inlet and outlet. This enthalpy change can be calculated by

$$\dot{Q}_{\text{steady-state, in}} = \Delta\dot{H}_{\text{oil}} = \dot{m}_{\text{oil}}\; c_{\text{oil}}\,(T_{\text{in}} - T_{\text{out}}) \tag{6.2}$$

with the mass flow rate $\dot{m}_{\text{oil}}$ and heat capacity c_{oil} of the thermal oil.

To achieve steady state, the adsorption-TES unit is heated to a constant temperature for at least 2 hours. We did this for three different temperatures of the adsorber inlet from 150 °C to 250 °C. The resulting temperatures at the measurement positions in the adsorber (cf. Figure 5.2) are shown in Figure 6.1. To analyze the influence of the temperature in the evaporator on the results, the steady-state measurements are repeated for all adsorber temperatures with 3 evaporator temperatures of 40 °C, 60 °C and 80 °C. Table 6.1 gives the heat loss of the storage unit during steady state $Q_{\text{steady-state, loss}}$ for the temperatures of the adsorber inlet. With rising temperature in the adsorber, the heat losses increase as expected.

The uncertainties of measured quantities in this chapter are determined as expanded measurement uncertainty $U^{95\%}$ with a coverage probability of 95 % according to the *Guide*

Table 6.1: Heat loss of the adsorber during steady state $Q_{\text{steady-state, loss}}$ and corresponding expanded, relative measurement uncertainty $U^{95\%}_{\text{rel},Q_{\text{loss}}}$ for varying temperatures at the adsorber inlet T_{in}.

T_{in} in °C	$Q_{\text{steady-state, loss}}$ in W	$U^{95\%}_{\text{rel},Q_{\text{loss}}}$ in %
250	252	16
200	171	21
150	105	27

to the Expression of Uncertainty in Measurement (GUM) [132]. Error bars in plots show the expanded uncertainties $U^{95\%}$. Further details can be found in Appendix B.

The expanded measurement uncertainty $U^{95\%}_{\dot{Q}_{\text{steady-state, loss}}}$ of the steady-state heat loss is 28–41 W with rising temperature. The resulting relative measurement uncertainty $U^{95\%}_{\text{rel},Q_{\text{loss}}}$ of the steady-state heat loss is given in Table 6.1.

6.1.1 Heat transfer from adsorber casing to environment

The steady-state measurements allow for a quantification of the temperature-dependent heat losses of the adsorption-TES unit. We thus derive a temperature-dependent heat-loss coefficient to estimate the heat losses during cyclic operation of the storage unit. During cyclic operation, heat-transfer regimes might differ from steady state, since the water vapor induces convective heat flow inside the adsorber casing. In contrast, outside the vacuum chamber, the heat-transfer mechanisms depend only on the temperature of the casing. Hence, we consider the heat transfer outside the adsorber casing: we calculate

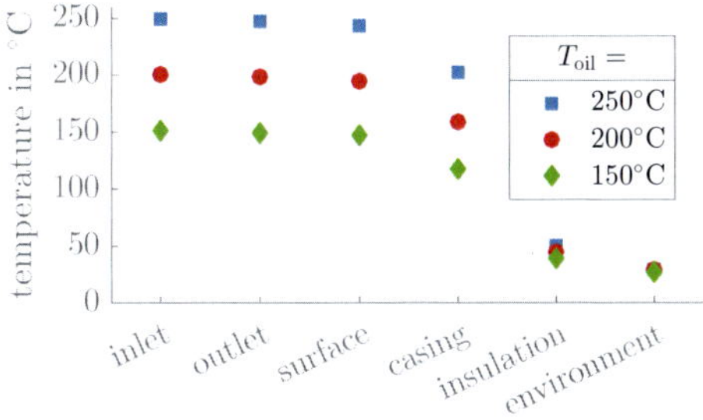

Figure 6.1: Steady-state temperatures at the measurement positions defined in Figure 5.2 for T_{in} = 150–250 °C at the adsorber inlet and T_{evap} = 40 °C in the evaporator.

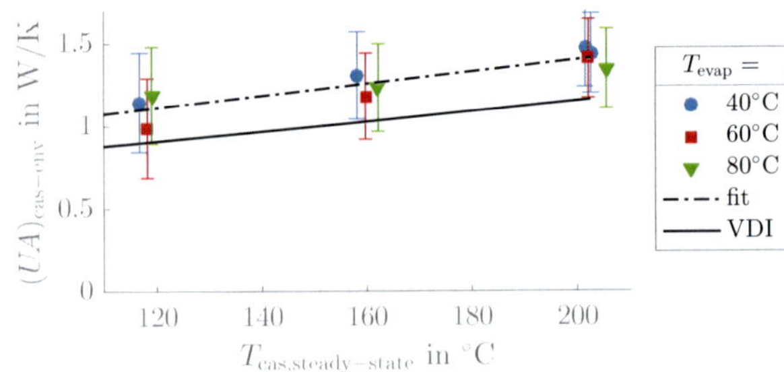

Figure 6.2: Heat-loss coefficients $(UA)_{\text{cas-env}}$ from adsorber casing to environment as a function of temperature of the adsorber casing, calculated from steady-state measurements (cf. Equation (6.3)) with $T_{\text{evap}} = 40, 60, 80\,°\text{C}$, with expanded uncertainty $U^{95\%}_{(\text{UA})}$, and $(UA)_{\text{cas-env,VDI}}$ according to Equation (6.4).

the temperature-dependent heat-loss coefficient $(UA)_{\text{cas-env}}$ from the adsorber casing to the environment, using the measured heat losses during steady-state measurements $\dot{Q}_{\text{steady-state, loss}}$ (Equation (6.2)) and the corresponding temperature of the adsorber casing T_{cas} and ambient temperature T_{env} (cf. Figure 6.1):

$$(UA)_{\text{cas-env}} = \frac{\dot{Q}_{\text{steady-state, loss}}}{T_{\text{cas}} - T_{\text{env}}} \tag{6.3}$$

Figure 6.2 shows the calculated heat-loss coefficients $(UA)_{\text{cas-env}}$ from the adsorber casing to the environment as a function of temperature of the adsorber casing. The increasing trend of the heat-loss coefficient with temperature is in accordance to the findings of Mawire et al. [84]. For different vaporization temperatures T_{evap}, the variation of the measured heat-loss coefficient $(UA)_{\text{cas-env}}$ is within the measurement uncertainty (cf. Figure 6.2). Accordingly, an influence of the evaporator temperature on the heat-loss coefficient $(UA)_{\text{cas-env}}$ can be neglected.

The relative uncertainty of the heat-loss coefficient $U^{95\%}_{\text{rel},(UA)_{\text{cas-env}}}$ is only half a percent higher than the relative uncertainty of the heat loss $U^{95\%}_{\text{rel},Q_{\text{loss}}}$ (Table 6.1). Due to the large temperature difference $(T_{\text{cas}} - T_{\text{env}})$ (cf. Figure 6.1), the uncertainties of the temperature-measuring devices are insignificant for the relative uncertainty of the heat-loss coefficient $U^{95\%}_{\text{rel},(UA)_{\text{cas-env}}}$. Yet, determination of the heat-loss coefficient for $T_{\text{in}} \leq 100\,°\text{C}$ is rather inaccurate due to the low heat flow rate values (Table 6.1). Hence, we only consider steady-state measurements at $T_{\text{in}} = 150$–$250\,°\text{C}$.

To confirm the physical relevance of our results, we compare the measured heat-loss coefficients $(UA)_{\text{cas-env}}$ to (UA)-values calculated with Nusselt correlations for heat transfer from a cylinder wall through insulation to the environment according to the VDI heat atlas [122]:

$$(UA)_{\text{cas-env,VDI}} = \left(\underbrace{\frac{d_{\text{ins}}}{\lambda_{\text{ins}}}(A_{\text{cas}} - A_{\text{ins}}) \ln(\frac{A_{\text{ins}}}{A_{\text{cas}}})}_{\text{conductive}} + \underbrace{\frac{1}{\alpha_{\text{ins}} A_{\text{ins}}}}_{\text{convective}} \right)^{-1} \tag{6.4}$$

Equation (6.4) includes a conductive term for the thickness of the rock wool insulation d_{ins}, the surface areas of the adsorber casing A_{cas} and of the adsorber insulation A_{ins}. The thermal conductivity of the insulation λ_{ins} is considered temperature-dependent according to the manufacturer [133]

$$\lambda_{\text{ins}} = \left(0.0363 + 0.0002\ T_{\text{cas}} + 5 \cdot 10^{-7}\ {T_{\text{cas}}}^2\right) \text{W}/(\text{m K}) \tag{6.5}$$

with T_{cas} in °C. The convective term in Equation (6.4) contains the heat-transfer coefficient α_{ins} on the outer surface of the adsorber insulation

$$\alpha_{\text{ins}} = \frac{\text{Nu}\ \lambda_{\text{air}}}{h_{\text{ins}}} \tag{6.6}$$

with the height of the insulated adsorber h_{ins} and the thermal conductivity of air λ_{air}. The Nusselt number Nu for free convection at the outer surface of the adsorber insulation is calculated by

$$\text{Nu} = \left(0.825 + 0.387\ (\text{Gr}_{\text{air}}\text{Pr}_{\text{air}} f_1)^{1/6}\right)^2 + 0.435\ \frac{h_{\text{ins}}}{d_{\text{ins}}} \tag{6.7}$$

according to the equation of Churchill and Chu [134] for a vertical plate with

$$f_1 = (1 + (0.492/\text{Pr}_{\text{air}})^{9/16})^{-16/9} \tag{6.8}$$

found by Churchill and Usagi [135]. The correction for a vertical cylinder ($0.435\ \frac{h_{\text{ins}}}{d_{\text{ins}}}$) was added by Fujii and Uehara [136]. The properties of air, like the Grashof number Gr and Prandtl number Pr, are calculated at ambient temperature.

Figure 6.2 reveals the qualitative agreement between the measured heat-loss coefficients $(UA)_{\text{cas-env}}$ (Equation (6.3)) and $(UA)_{\text{cas-env,VDI}}$ calculated by empirical Nusselt correlations (Equation (6.4)): the temperature dependence of the (UA)-value can be verified and the absolute values of $(UA)_{\text{cas-env}}$ and $(UA)_{\text{cas-env,VDI}}$ are in the same range. The measured heat-loss coefficient $(UA)_{\text{cas-env}}$ is 24 % higher than the empirical prediction. This is presumably due to convection in the room around the storage unit. Convection is induced by radiator heat sources in the room and a cold window-front. The stronger convection in the room is not predicted by the Nusselt correlation for free convection (cf. Equation (6.7)).

Since the temperature-dependent heat-loss regime outside the adsorber casing should be the same during steady-state and cyclic operation, we use our calculated heat-loss coefficients $(UA)_{\text{cas-env}}$ to determine the heat losses during storage cycles. We therefore fit a linear temperature-dependent heat-loss coefficient $(UA)_{\text{cas-env}}$ to our measurement data. The fit is shown in Figure 6.2.

6.1.2 Radiative heat losses inside the adsorber casing

Dawoud et al. [35] and Li et al. [33] emphasize the importance of radiation for heat losses of closed adsorption-TES units. We therefore calculate radiative losses inside the casing by

$$\dot{Q}_{\text{radiation}} = \sigma \frac{A_{\text{ads}}}{1/\varepsilon_{\text{ads}} + A_{\text{ads}}/A_{\text{cas}}\left(1/\varepsilon_{\text{cas}} - 1\right)} \left(T_{\text{ads}}^4 - T_{\text{cas}}^4\right) \tag{6.9}$$

with the Stefan-Boltzmann-constant $\sigma = 5.67\,\text{W/m}^2/\text{K}^4$, the emissivities of the adsorber casing ε_{cas} and of the adsorber surface ε_{ads} and with the measured temperatures of the adsorber surface T_{ads} and casing T_{cas}.

We estimate the emissivity of the adsorber casing $\varepsilon_{\text{cas}} = 0.2$ with an infrared camera by heating the casing material to temperatures in the range of $75 - 180\,°\text{C}$. For the emissivity of the adsorber surface, we assume $\varepsilon_{\text{ads}} = 0.89$ which is a typical value for zeolite [137]. Since the perforated sheet at the adsorber surface is made of steel with a lower emissivity, this assumption probably leads to an overestimation of the radiative heat losses.

The measured heat losses for steady-state conditions and the radiative losses inside the casing are shown in Figure 6.3. The share of radiation in total heat loss is nearly constant with the temperature of the adsorber outlet $T_{\text{out,steady-state}}$:

$$\frac{\dot{Q}_{\text{radiation}}}{\dot{Q}_{\text{steady-state, loss}}} \approx 57\,\%. \tag{6.10}$$

Besides radiation, convection and thermal bridges contribute to the total heat flow rate inside the casing. A radiation shield, as proposed by Dawoud et al. [35], could reduce the

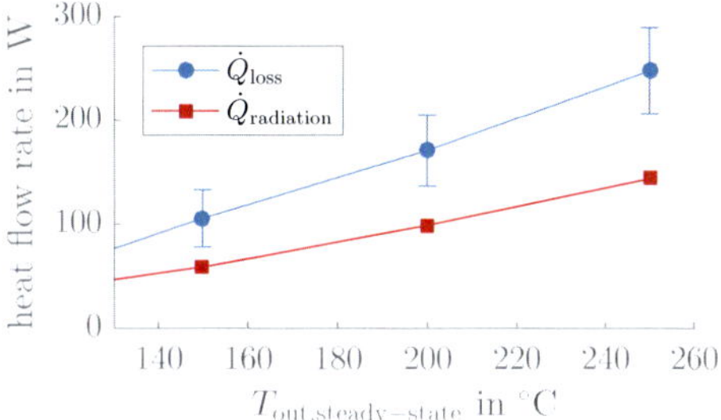

Figure 6.3: Heat losses from steady-state measurements at $T_{evap} = 40\,°C$: measured total heat loss $\dot{Q}_{steady-state,\,loss}$ (cf. Equations (6.2) and (6.1)) with expanded uncertainty as error bars and radiative heat flow rate $\dot{Q}_{radiation}$ from adsorber to casing calculated according to Equation (6.9).

loss, but no more than 57 %. The share of radiation in total heat loss could even decrease during storage cycles, if convection increases due to water vapor flow.

6.2 Storage-cycle measurements

The analysis of storage cycles is crucial to investigate the performance of a thermal energy storage (TES) unit. A complete storage cycle includes charging (desorption) and discharging (adsorption) and sometimes a storage period in between. In Section 6.2.1, we present measurements with direct discharge without a storage period. We derive the heat losses during charging and discharging, the resulting efficiencies and the total energy recovery ratio (ERR) of our adsorption-TES unit (Section 6.2.2). In Section 6.2.3, we investigate measurements of cycles with 2 hours storage time between charging and discharging. For comparison, a cycle measurement is also presented for seasonal storage.

In our analysis, we vary the maximum temperature at the end of charging in the adsorber from 175 up to 250 °C (T_{des} = 175/200/225/250 °C), because zeolite is nearly dry at 250 °C [85]. Potential heat sources may be the exhaust gas from cogeneration, surplus energy from renewable eletricity production or solar thermal collectors. Even though recent solar thermal technologies are restricted to temperatures up to 180 °C [138], Li et al. [139] expect development potential for solar systems to supply heat at temperatures up to 250 °C.

All other temperatures which define the storage process are constant: the temperature at the end of discharging is $T_{\mathrm{ads}} = 70\,^{\circ}\mathrm{C}$ and the evaporator is kept constant at $T_{\mathrm{evap}} = T_{\mathrm{cond}} = 40\,^{\circ}\mathrm{C}$ both during charging, when the desorbed water is condensed, and during the vaporization process during discharging. The discharging temperature is consistent with the heat supply temperature of a conventional radiator heating system [140].

The storage-cycle measurement procedure is as follows: the charging period is terminated when the temperature of the adsorber outlet T_{out} reaches the maximum desorption temperature T_{des}. Discharging is induced by cooling of the heat-transfer oil. As soon as the temperature of the adsorber outlet has reduced to $T_{\mathrm{out}} = T_{\mathrm{ads}} = 70\,^{\circ}\mathrm{C}$, we terminate discharging and start with the next charging period. It is important to note that this operation procedure leads to incomplete de- and adsorption, i.e. the potential of the storage material is not fully explored. We chose this procedure, since it represents more realistic performance than waiting until the heat flow rate approaches zero and equilibrium is reached in the adsorber.

For each temperature set, we measured at least three successive storage cycles. The evaluation starts with the second successive cycle to ensure reproducible starting conditions. The heat flow rate transferred to the adsorber during cycle measurements is calculated by

$$\dot{Q}_{\mathrm{cycle,in}} = \Delta\dot{H}_{\mathrm{oil}} - \frac{\mathrm{d}U_{\mathrm{oil}}}{dt}\,. \tag{6.11}$$

The enthalpy difference of the heat-transfer oil $\Delta\dot{H}_{\mathrm{oil}}$ is reduced by the change of internal energy of the oil U_{oil} inside the adsorber heat exchanger.

$\Delta\dot{H}_{\mathrm{oil}}$ is calculated according to Equation (6.2) with data from cycle measurements (Figure 6.4 and Figure 6.6 top). The change of internal energy of the heat-transfer oil is given by

$$\frac{\mathrm{d}U_{\mathrm{oil}}}{dt} = \varrho_{\mathrm{oil}} V_{\mathrm{oil}} c_{\mathrm{oil}} \frac{\mathrm{d}T_{\mathrm{oil}}}{dt} \tag{6.12}$$

with the density ϱ_{oil}, volume V_{oil} and heat capacity c_{oil} of the oil. We thereby neglect an influence of oil pressure and volume change on the internal energy leading to an error of less than 0.1 %.

During storage-cycle measurements, the heat losses of the storage unit are calculated with the (UA)-values from the steady-state measurements (cf. Section 6.1.1) and with the measured temperatures of the adsorber casing T_{cas} and of the environment T_{env} from the storage-cycle measurements:

$$\dot{Q}_{\mathrm{cycle\ loss}} = (UA)_{\mathrm{cas-env}}\,(T_{\mathrm{cas}} - T_{\mathrm{env}})\,. \tag{6.13}$$

6.2.1 Measurements with direct discharging

In a simple storage cycle, charging (desorption) is directly followed by discharging (adsorption) without delay. Figure 6.4 shows the temperatures (top) and heat flow rates (bottom) of a typical measurement for a maximum desorption temperature $T_{des} = 250\,°C$.

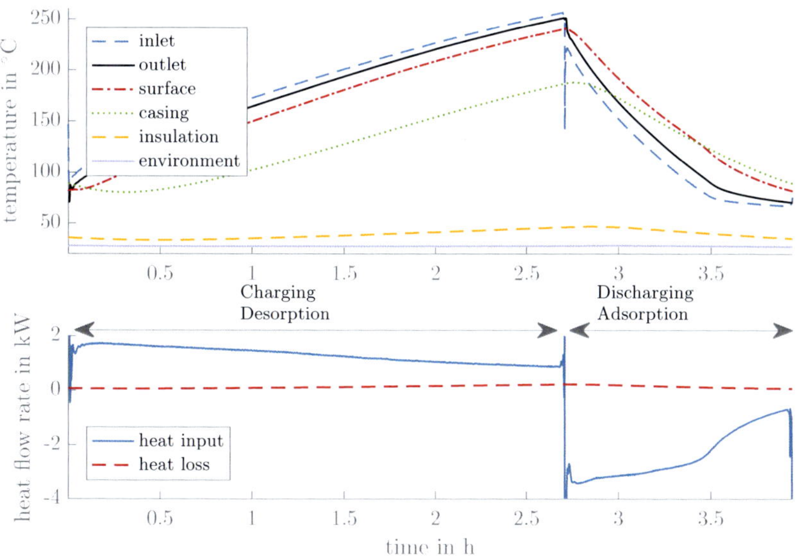

Figure 6.4: Storage-cycle measurement with direct discharge and maximum temperature of the adsorber outlet during desorption $T_{des} = 250\,°C$ and $T_{evap} = 40\,°C$ in the evaporator. Top: Measured ambient and adsorber temperatures at positions shown in Figure 5.2. Bottom: Adsorber heat input (Equation (6.11)) and adsorber heat loss (Equation (6.13)).

The heat input to the adsorber $\dot{Q}_{cycle,in,des}$ during charging is provided by the electric heater with constant power of 3 kW. Since the oil circulation system suffers from heat losses, $\dot{Q}_{cycle,in,des}$ drops with rising temperature from 1.7 kW to 0.9 kW (cf. Figure 6.4, bottom). During discharging, the adsorber heat flow rate $\dot{Q}_{cycle,in,ads}$ evolves from −3.5 kW to −0.7 kW (cf. Figure 6.4, bottom). A bypass-valve at the secondary heat exchanger controls the discharging heat flow rate. The heat flow rate is maintained constant as long as possible. As soon as the bypass-valve is fully open, the discharging heat flow rate reduces due to a decreasing temperature difference in the secondary heat exchanger. At this secondary heat exchanger, the outlet temperature is maintained at 65 °C for discharging of the adsorber.

The sharp peaks in the adsorber heat input occur when valves in the oil circulation system are switched: cold or hot oil coming from the electric heater or the secondary heat exchanger passes through the adsorber. As a result, the heat flow rate appears to rise for a short time.

6.2.2 Heat-loss analysis

The findings for adsorber heat losses from steady-state measurements (Section 6.1.1 and Section 6.1.2) are used to analyze the performance of the adsorption-TES unit during storage cycles.

The share of adsorber heat losses $\dot{Q}_{\text{cycle loss}}$ (Equation (6.13)) in the adsorber heat flow rate $\dot{Q}_{\text{cycle,in}}$ (Equation (6.11)) rises from 3.1 % at the beginning of a storage cycle to 24 % at the end of desorption. The heat loss share reaches its highest value for $T_{\text{des}} = 250\,°\text{C}$ (cf. Figure 6.4, bottom). The share of radiative heat losses $\dot{Q}_{\text{radiation}}$ in the adsorber heat flow rate $\dot{Q}_{\text{cycle,in}}$ rises from 0 % to 17 % at $T_{\text{des}} = 250\,°\text{C}$. Compared to the total heat losses during one complete storage cycle $Q_{\text{cycle loss}}$, the calculated heat losses due to radiation correspond to 57 % as observed for the steady state (Section 6.1.2). It has to be noted that radiative heat losses are calculated for a worst-case estimate with a very high adsorber emissivity (cf. Section 6.1.2).

To analyze the storage performance in terms of energy recovery ratio, we calculate the total heat input to the storage unit during charging

$$Q_{\text{des}} = \int_{t_{\text{des,start}}}^{t_{\text{des,end}}} \dot{Q}_{\text{cycle,in}}\, \mathrm{d}t \tag{6.14}$$

and the total heat output of the storage unit during discharging

$$Q_{\text{ads}} = \left| \int_{t_{\text{ads,start}}}^{t_{\text{ads,end}}} \dot{Q}_{\text{cycle,in}}\, \mathrm{d}t \right| . \tag{6.15}$$

The energy recovery ratio ERR is the ratio of heat input to heat output, cf. Equation (2.3)

$$\text{ERR} = \frac{Q_{\text{ads}}}{Q_{\text{des}}} . \tag{6.16}$$

This definition is equal to the energy efficiency in Reference [33] or to the COP_2 in Reference [35]. In Section 4.3, we refer to this ratio as $\eta_{\text{E+C-}}$ (cf. Table 4.2). The energy recovery ratio of storage cycles with direct discharge is referred to as $\mathrm{ERR}_{\mathrm{direct}}$ in the following.

The total heat losses during the processes of desorption and adsorption are calculated by

$$Q_{\mathrm{loss,\ des}} = \int_{t_{\mathrm{des,start}}}^{t_{\mathrm{des,end}}} \dot{Q}_{\mathrm{cycle,loss}}\,\mathrm{d}t \tag{6.17}$$

$$Q_{\mathrm{loss,\ ads}} = \int_{t_{\mathrm{ads,start}}}^{t_{\mathrm{ads,end}}} \dot{Q}_{\mathrm{cycle,loss}}\,\mathrm{d}t \tag{6.18}$$

with heat losses $\dot{Q}_{\mathrm{cycle,loss}}$ according to Equation (6.13) using the (UA)-values determined from steady-state measurements (cf. Section 6.1). The energy recovery ratio for direct discharge can also be calculated by

$$\mathrm{ERR}_{\mathrm{calc}} = \eta_{\mathrm{in}}\eta_{\mathrm{out}} \tag{6.19}$$

according to the definition by Zanganeh et al. [141] with efficiencies for charging

$$\eta_{\mathrm{in}} = Q_{\mathrm{stored,des}}/Q_{\mathrm{des}} \tag{6.20}$$

and discharging

$$\eta_{\mathrm{out}} = Q_{\mathrm{ads}}/Q_{\mathrm{stored,ads}} \; . \tag{6.21}$$

The charging efficiency η_{in} (Equation (6.20)) is the ratio of the total thermal energy which is stored after the desorption process

$$Q_{\mathrm{stored,des}} = Q_{\mathrm{des}} - Q_{\mathrm{loss,des}} \tag{6.22}$$

to the total heat input to the storage unit Q_{des} (Equation (6.14)). The discharging efficiency η_{out} (Equation (6.21)) is the ratio of the total heat output of the storage unit Q_{ads} (Equation (6.15)) to the total thermal energy in the storage unit before the adsorption process starts

$$Q_{\mathrm{stored,ads}} = Q_{\mathrm{ads}} + Q_{\mathrm{loss,ads}} \; . \tag{6.23}$$

The results of the storage-cycle measurements with direct discharge are shown in Figure 6.5. The measured energy recovery ratio for direct discharge $\mathrm{ERR}_{\mathrm{direct}}$ rises from 85 to 91 % with decreasing temperature of desorption T_{des}. The discharging efficiency η_{out} remains high (95-96 %) while the charging efficiency η_{in} drops from 96 to 91 % for higher temperatures of desorption. This reduction of charging efficiency is due to increasing heat losses. First, heat losses increase due to a higher temperature difference between the adsorber and the environment. Second, the charging heat flow rate reduces with rising temperature in the oil

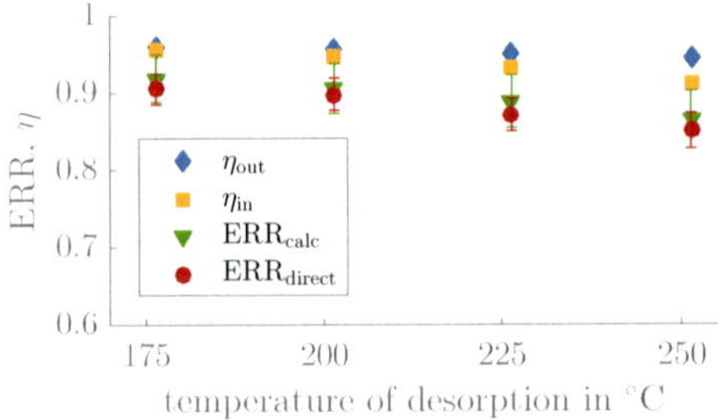

Figure 6.5: Energy recovery ratios $\mathrm{ERR_{direct}}$ (Equation (6.16)) and $\mathrm{ERR_{calc}}$ (Equation (6.19)) (with error bars) as well as charging efficiencies η_{in} (Equation (6.20)) and discharging efficiencies η_{out} (Equation (6.21)) for direct discharge measurements at $T_{\mathrm{des}} = 175$–$250\,°\mathrm{C}$.

circulation system (cf. Figure 6.4). Thus, it takes longer until the maximum temperature in the adsorber T_{des} is reached and more heat is lost.

In general, the quantitative results of the energy recovery ratio ERR depend on the heat flow rate during charging and discharging: with a lower charging rate, the storage temperature increases more slowly and it thus takes longer to achieve the final charging state, i.e. the desired maximum temperature. The effect is similar for the discharging process: lower heat flow rates leave the storage unit at higher temperature for a longer period of time. Hence, slower charging or discharging leads to larger heat losses, at least for the same maximum storage temperature. As a consequence, the results from the simulation study in Section 4.3, with a charging period of 5 h, are not comparable to the measurements in this chapter with approximately 2.5 h of charging (cf. Figure 6.4).

The efficiencies of the adsorption-TES unit are obtained on the basis of heat losses calculated with the heat-loss coefficient $(UA)_{\mathrm{cas-env}}$ from steady-state measurements (Section 6.1.1). To validate this coefficient, we compared the calculated and the measured energy recovery ratios and found that $\mathrm{ERR_{direct}} \approx \mathrm{ERR_{calc}} = \eta_{\mathrm{in}}\eta_{\mathrm{out}}$. $\mathrm{ERR_{calc}}$ is always slightly higher than $\mathrm{ERR_{direct}}$, but within the measurement uncertainty (Figure 6.5). This difference in energy recovery ratio is probably due to a certain loss of stored energy during the switching process between de- and adsorption. The maximum relative measurement uncertainty of $\mathrm{ERR_{direct}}$ is $U^{95\%}_{\mathrm{rel,ERR_{direct}}} = 2.3\,\%$ and for the calculated $\mathrm{ERR_{calc}}$, it is $U^{95\%}_{\mathrm{rel,ERR_{calc}}} = 3.8\,\%$.

6.2.3 Measurements with 2 hours storage time

Thermal energy storage is often applied to processes where heat supply and demand are separated in time. To investigate the influence of such a storage period on the performance of our adsorption-TES unit, we chose the following procedure: charging (desorption) is followed by 2 hours storage time where the valve is closed between adsorber and evaporator. Afterwards, we initiate discharging (adsorption) by opening the valve 30 s after the oil circulation pump is turned on.

Figure 6.6 shows the temperatures (top) and the heat flow rates (bottom) of a typical measurement with a maximum desorption temperature $T_{des} = 250\,°C$ and 2 hours storage time. During storage, the temperatures in the oil circulation system decrease faster than

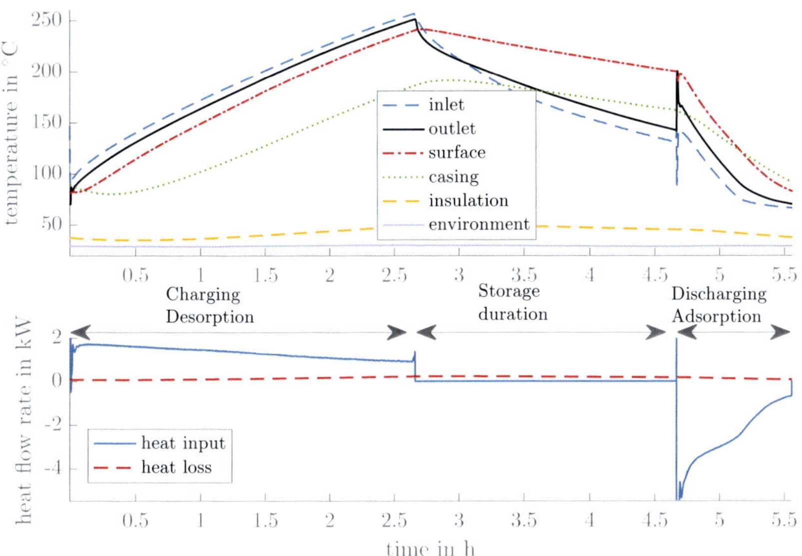

Figure 6.6: Storage-cycle measurement with 2 hours storage time with maximum temperature in the adsorber outlet $T_{des} = 250\,°C$ and $T_{evap} = 40\,°C$ in the evaporator. Top: Measured ambient and adsorber temperatures at positions shown in Figure 5.2. Bottom: Heat input to the adsorber (Equation (6.11)) and heat losses (Equation (6.13)).

the temperature of the adsorber surface (Figure 6.6 top), because the adsorber is much better insulated than the oil pipes.

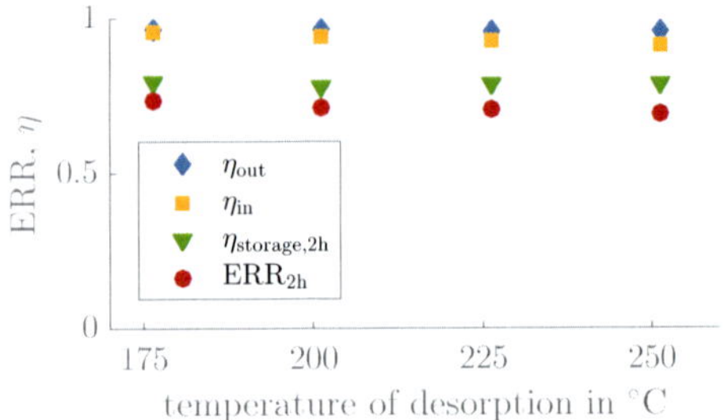

Figure 6.7: Measured energy recovery ratio (ERR) (Equation (6.16)) and calculated efficiencies η_{in}, η_{out} and η_{storage} (Equations (6.20), (6.21) and (6.25)) for measurements with 2 hours storage time at different desorption temperatures T_{des}.

The performance of the storage unit is quantified by the energy recovery ratio $\text{ERR}_{2\text{h}}$ which is plotted in Figure 6.7. The $\text{ERR}_{2\text{h}}$ drops to 69–74 %. This is due to increased heat losses compared to operation with direct discharge ($\text{ERR}_{\text{direct}} = 85\text{-}91\,\%$, cf. Section 6.2.1). The result for the charging and discharging efficiencies are nearly the same as for operation with direct discharge: the efficiency of the charging process is $\eta_{\text{in}} = 92\text{–}96\,\%$. The discharging efficiency is constant $\eta_{\text{out}} = 96\,\%$.

In general, for storage cycles with a storage period, the energy recovery ratio becomes

$$\text{ERR}_{2\text{h}} = \eta_{\text{in}}\eta_{\text{storage}}\eta_{\text{out}} \tag{6.24}$$

with η_{storage} as the ratio of stored thermal energy before discharging $Q_{\text{stored,ads}}$ (Equation (6.23)) to stored thermal energy after charging $Q_{\text{stored,des}}$ (Equation (6.22)). We calculate the efficiency of the storage period η_{storage}

$$\eta_{\text{storage}} = \text{ERR}_{2\text{h}} / (\eta_{\text{in}}\eta_{\text{out}}) \tag{6.25}$$

by using the definition of $\text{ERR}_{2\text{h}}$ from Equation (6.16).

The efficiency of the storage period η_{storage} for 2 hours storage time is constant around 79 %. We reckon that the largest part of heat losses occur during the storage period when the temperature inside the storage unit is at a high level for a long time (cf. Figure 6.6, temperature of the adsorber surface).

6.2.4 Seasonal-storage measurement

To complete the experimental analysis of our adsorption-TES unit, we specify a measurement procedure to emulate seasonal storage with a maximum desorption temperature of $T_{\text{des}} = 250\,°\text{C}$. The charging process is again accomplished by heating the adsorber module until the outlet temperature reaches 250 °C. Then, we turn off all pumps, close the valve between adsorber and evaporator and wait until the complete storage unit is at ambient temperature. The cooldown to ambient temperature takes one week.

The discharging is prepared by heating the evaporator to 40 °C. Then, the valve to the adsorber is opened. First, we wait until the temperature of the adsorber surface exceeds 70 °C, then we turn on the oil pump to recover the discharging heat. To be comparable to the cyclic measurements before, we terminate the adsorption process when the temperature in the adsorber outlet falls below $T_{\text{ads}} = 70\,°\text{C}$ again.

During cooldown to ambient temperature all sensible heat is lost. The results of the seasonal measurements thus provide a lower bound for the energy recovery ratio of our storage unit: $\text{ERR}_{\text{seasonal}} = 14\,\%$. The metal parts in the storage unit do not contribute to adsorption TES. The adsorber heat exchanger is designed with a large share of metal lamellae compared to adsorbent to ensure high heat transfer rates (cf. Section 3.1). All this metal has to be heated up for discharging. This heat has to be provided by the storage unit itself lowering the energy that can be recovered. For measurements with lower charging temperatures than 250 °C, the storage unit hardly provides heat at discharging temperatures above 70 °C. Thus, lower temperatures of discharging would be beneficial.

6.3 Assessment of storage performance

The energy recovery ratio (ERR) evidently rises when heat losses decrease at lower desorption temperatures (cf. Section 6.2). However, higher desorption temperatures lead to an increase of the amount of stored thermal energy. As a result, the performance of a thermal energy storage (TES) unit is a compromise between the energy recovery ratio ERR and the energy storage density ESD. This section provides a comprehensive analysis of the ERR-ESD trade-off.

We define the energy storage density ESD (cf. Equation (2.1)) as ratio of the discharging heat Q_{ads} (Equation (6.15)) and the total storage unit volume V_{store}

$$\text{ESD} = \frac{Q_{\text{ads}}}{V_{\text{store}}} \,. \tag{6.26}$$

The total volume V_{store} includes adsorber, evaporator and insulation (cf. Section 5.2) to account for the total space requirements of the storage unit. As we already discussed in Section 2.2, the reference volume has to be considered when comparing results for the energy storage density from literature. Values of energy storage densities are often given based on the volume or mass of adsorbent, as by Li et al. [33] or based on the volume of the adsorber, as by Dawoud et al. [35].

We define the energy storage density ESD with the total heat recovered during discharging Q_{ads}, because we consider this definition (Equation (6.26)) of the energy storage density a reasonable measure of the real potential of a storage unit to provide heat. Alternatively, we could have based our analysis on the maximum chemical potential of the material or on the charging heat. However, we regard ESD values from maximum potential or charging heat as misleading for practitioners, since they neglect heat losses. The energy storage density ESD should thus be calculated with the heat output, as in References [32, 33, 35].

Lossless operation rather provides an upper limit for the energy storage density. If the charging process was lossless, the total thermal energy in the storage unit after the desorption process $\text{Q}_{\text{stored,des}}$ (Equation (6.22)) would be the stored thermal energy. This amount of thermal energy could then be recovered in a lossless discharging process. We hence define

$$\text{ESD}_{\text{lossless}} = \frac{\text{Q}_{\text{stored,des}}}{\text{V}_{\text{store}}} \,, \tag{6.27}$$

according to the definition of the energy storage density in Equation (6.26). The energy recovery ratio would be $\text{ERR}_{\text{lossless}} = 1$ for lossless storage.

Figure 6.8 compares the performance for lossless operation to our measurements with direct discharge, with 2 hours storage time and after cooldown to ambient temperature (cf. Section 6.2). The results reveal the trade-off between energy storage density ESD and energy recovery ratio ERR: with increasing temperature of desorption T_{des}, more energy is stored; at the same time, heat losses reduce the energy recovery ratio. However, for storage cycles with equal storage time, ESD rises significantly with rising temperature while ERR drops only slightly with rising temperature. For cycles with direct discharge, $\text{ERR}_{\text{direct}}$ remains higher than 85 % for all temperatures while $\text{ESD}_{\text{direct}}$ nearly doubles from 11.1 kW h/m^3 at 175 °C to 20.4 kW h/m^3 at 250 °C. Yet, with longer storage time, ERR

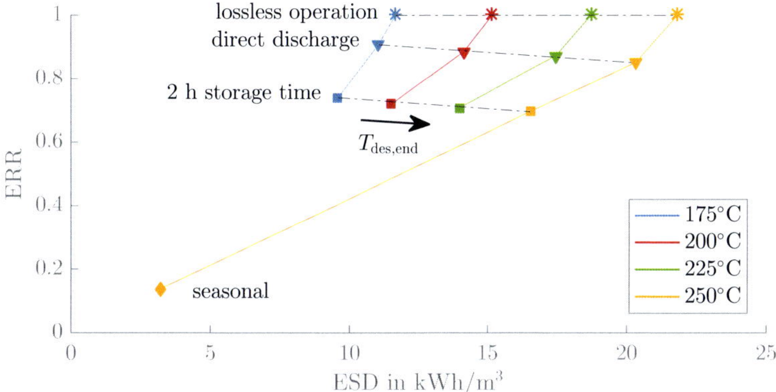

Figure 6.8: Energy storage density ESD (Equation (6.26)) versus energy recovery ratio ERR (Equation (6.16)) for desorption temperatures $T_{\text{des}} = 175$–$250\,°\text{C}$ for measurements with direct discharge, measurements with a storage time of 2 hours and a measurement with discharging after cooldown to ambient temperature (seasonal storage, cf. Section 6.2.4). For comparison: upper limit for lossless operation $\text{ESD}_{\text{lossless}}$ (Equation (6.27)) (lines just guide the eye).

significantly decreases, e.g. for 2 hours storage time, to $\text{ERR}_{\text{2h}} < 74\,\%$. The storage density decreases as well: $\text{ESD}_{\text{2h}} = 9.5$–$16.6\,\text{kW h/m}^3$. After cooling the storage unit to ambient temperature, part of the stored thermal energy can still be recovered: $\text{ERR}_{\text{seasonal}} = 14\,\%$ and $\text{ESD}_{\text{seasonal}} = 3.3\,\text{kW h/m}^3$. Thus, even small-scale seasonal storage is possible, though all sensible heat is lost and part of the adsorption enthalpy has to be used to reheat the system.

We compare the energy storage densities of our adsorption TES system to the upper limit for lossless operation at the respective temperature (cf. Figure 6.8). The ratio $\text{ESD}/\text{ESD}_{\text{lossless}}$ is independent of the desorption temperature: for direct discharge $\text{ESD}_{\text{direct}}$ is 93 % of the upper limit and for 2 hours storage time ESD_{2h} is 76 % of the upper limit.

The energy storage density could be enhanced even further with an increased zeolite to metal ratio. In particular, the energy storage density for seasonal storage $\text{ESD}_{\text{seasonal}}$ should increase with less metal in the adsorber module. However, the heat flow rate would be limited due to poorer heat transfer into the adsorbent bed. For a seasonal storage application, the storage unit would also need to be larger. The investigated storage unit is suited for short-term storage rather than for long-term storage.

To compare our results with other authors, we convert the energy storage density ESD in Figure 6.8 to an energy storage density $\mathrm{ESD_{HX}}$ based on the adsorber heat exchanger volume V_{HX} by multiplying the ESD (Equation (6.26)) with $V_{\mathrm{store}}/V_{\mathrm{HX}} = 6.67$:

$$\mathrm{ESD_{HX}} = \frac{Q_{\mathrm{ads}}}{V_{\mathrm{HX}}}. \tag{6.28}$$

Exemplarily, for desorption to 250 °C and direct discharge, the energy storage density is $\mathrm{ESD_{HX}} = 136\,\mathrm{kW\,h/m^3}$. When applying Equation (6.28) to the results of Dawoud et al. [35], the energy storage density is $56\,\mathrm{kW\,h/m^3}$ based on their adsorber heat exchanger volume. For the adsorber of Li et al. [33], the energy storage density is $39\,\mathrm{kW\,h/m^3}$ based on their adsorber heat exchanger volume. It is important to note that the temperatures and storage times have to be considered as well when comparing values for the energy storage density.

6.4 Summary

In this chapter, heat losses are experimentally analyzed for an adsorption-TES unit. To determine the heat-loss coefficients, we use steady-state measurements. We further show that the heat losses are independent of the evaporator temperature T_{evap}.

For cycles with direct discharge and cycles with 2 hours storage time, we quantify the thermal efficiencies of our storage unit. For both series of measurements, we obtain similar charging ($\eta_{\mathrm{in}} > 91\,\%$) and discharging efficiencies ($\eta_{\mathrm{out}} > 95\,\%$). Most heat losses occur during the storage period, where the efficiency for 2 h storage time is $\eta_{\mathrm{storage}} = 79\,\%$.

Our comprehensive analysis of the storage performance reveals a trade-off between the energy storage density and the energy recovery ratio. The energy recovery ratio decreases with rising charging temperature from 175 to 250 °C: 91–85 % for direct discharge and 74–69 % with 2 h storage time. The energy storage density increases significantly from 9.5 to $20.4\,\mathrm{kW\,h/m^3}$ based on the storage unit volume, with rising charging temperature from 175 to 250 °C. The energy storage density for direct discharge reaches 93 % compared to the upper limit of lossless operation. For 2 h storage time, the energy storage density is around 76 % compared to lossless operation.

We also prove that long-term storage is feasible with our small-scale adsorption-TES unit by experimentally verifying seasonal storage with a charging temperature of 250 °C and discharging at 70 °C. The storage unit reaches an energy recovery ratio of 14 % and an energy storage density of 3.3 kW h/m^3. These values are significantly lower than for short-term storage, since only a small part of the stored thermal energy, i.e. the adsorption enthalpy, can be recovered after cooling to ambient temperature.

Our results show that heat losses cannot be neglected for adsorption TES. Despite the thermochemical effect through the adsorption enthalpy, sensible heat losses are crucial and reduce the storage performance. Nevertheless, heat losses of a short-term adsorption-TES unit are manageable: the energy recovery ratio remains higher than 85 % for direct discharge and higher than 69 % for short storage times.

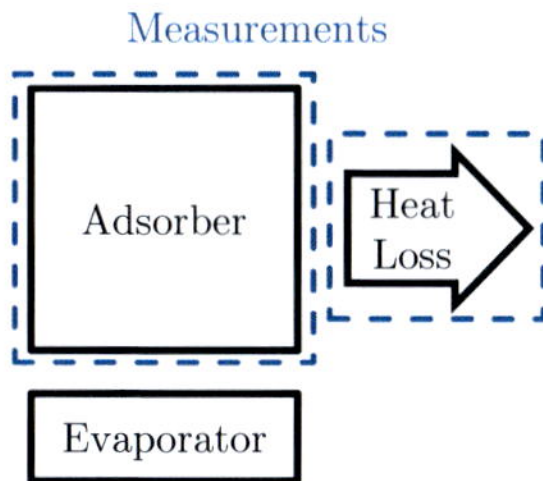

Figure 6.9: Measurements in Chapter 6: heat-loss coefficients are derived from experiments and the adsorber performance is experimentally analyzed.

To sum up, the heat-loss coefficients are now determined from measurements and the adsorber performance is thoroughly analyzed, cf. Figure 6.9. In the following Chapter 7, the heat-loss coefficients and the storage cycle measurements are used to develop a mathematical model that accurately describes the performance of the adsorption-TES unit including adsorber, evaporator and heat losses.

7 A Validated Model of the Adsorption Thermal Energy Storage Unit

In the previous Chapter 6, we derive heat-loss coefficients for our closed-adsorption thermal energy storage (adsorption TES) unit from steady-state measurements. Based on these findings, in this chapter, we improve our dynamic model from Chapter 4 and provide a thorough calibration and validation of the storage-unit model including adsorber, evaporator and heat losses (cf. Figure 7.1 in comparison to Figure 4.4).

The goal of this thesis is to provide an accurate, yet simple model to describe the behavior of our adsorption-TES unit for simulations on the energy-system level. For this purpose, we choose a dynamic lumped-parameter approach with one dimensional (1-D) discretization of the heat exchangers to describe the storage unit. Since adsorption TES is suitable for a wide range of applications, we test the model accuracy for analysis of storage cycles in both industrial applications such as brewing (cf. Chapter 4) as well as residential heating. The validated model thus provides a sound basis for future investigation on adsorption-TES systems.

This chapter starts with the introduction of the model setup in Section 7.1, focusing particularly on the changes and improvements compared to the previous model described

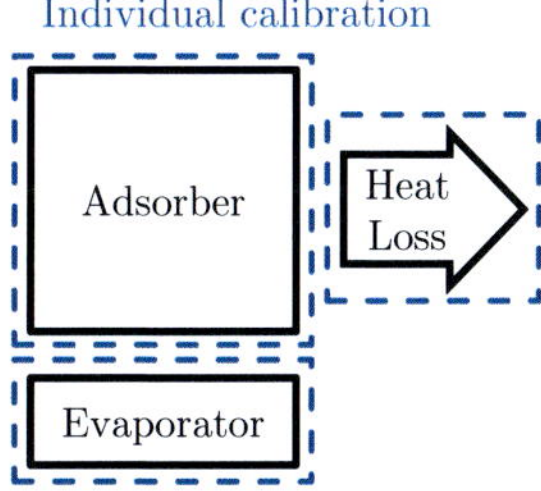

Figure 7.1: Identification of model parameters in Chapter 7: each part of the storage-unit model, i.e. adsorber, evaporator and heat losses, is individually calibrated with measurement data.

in Section 4.2.1. This improved model is then calibrated in Section 7.2. A simple storage cycle with process conditions of a residential heating application with direct discharge is used to calibrate the model, i.e. to determine the unknown model parameters. The simulation accuracy is quantified for the calibration measurement. Further measurements are introduced in Section 7.3 to validate the model and to quantify the prediction accuracy of the calibrated model in Section 7.4. The chapter is summarized in Section 7.5.

7.1 Improved model setup

The mathematical model of the adsorption-TES unit with a valve between adsorber and evaporator (cf. Section 5.2) is built upon the model established in Section 4.2.1. Due to the valve and the new experiments, the model setup (cf. Figure 7.2) is slightly refined compared to the model used for the industrial application study in Chapter 4. The general

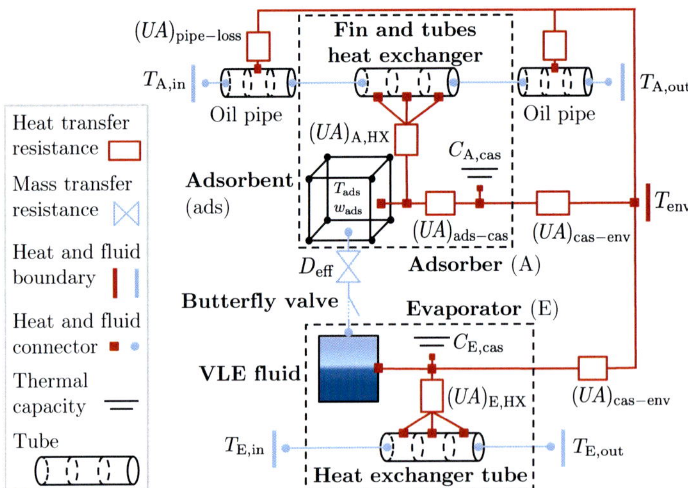

Figure 7.2: Model of adsorption thermal energy storage unit: adsorber (A) and evaporator (E) contain models of the adsorption working pair, thermal capacities (C) of casings (cas), 1-dimensional heat exchangers (HX). The butterfly valve is modeled as a flap and mass transfer resistance is described with the effective coefficient D_{eff}. Heat transfer resistances are modeled with the effective coefficients UA.

model assumptions and equations remain as described in Section 4.2.1. In contrast to Chapter 4, we improve the description of (1) the adsorption and thermophysical properties of the zeolite, (2) the heat transfer to the environment, (3) the evaporator and (4) the connection between adsorber and evaporator. These improvements are possible due to the enhanced experimental setup (cf. Chapter 5) and the thorough examination of heat losses in Chapter 6.

7.1.1 Adsorption and thermophysical properties of the zeolite

For the description of the adsorption process in the storage unit, the equilibrium data of zeolite 13 X and water is described according to Dubinin's theory of micropore filling [119]. For the improved model, the equilibrium data is adjusted to our own measurements (cf. Appendix A.1). Following Schawe [111], an arctangent function is used to describe the characteristic function of the equilibrium data of zeolite and water. The characteristic function gives the specific volume of adsorbate W as function of the adsorption potential A in equilibrium state:

$$W(A) = W_0 + \frac{W_1}{\pi}\left(\arctan(\frac{A - a_1}{a_2}) + \frac{\pi}{2}\right) \tag{7.1}$$

with the maximum adsorbate volume $W_0 = 0.2216\,\mathrm{l/kg}$, the coefficients $W_1 = -0.2152\,\mathrm{l/kg}$, $a_1 = 850\,\mathrm{kJ/kg}$ and $a_2 = 379.6\,\mathrm{kJ/kg}$ and with the adsorption potential A

$$A = RT\ln\left(\frac{p_s(T)}{p}\right) \tag{7.2}$$

depending on the adsorbent temperature T, the pressure p and the saturation pressure p_s at the respective temperature. The characteristic function which is implemented in the model and the measured equilibrium data are shown in Figure 7.3.

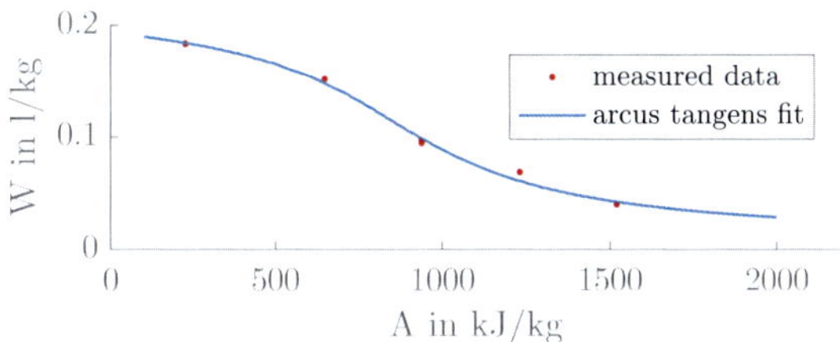

Figure 7.3: Characteristic function for zeolite 13 X.

Furthermore, to describe the thermophysical behavior of the zeolite in a large range of temperatures from environmental temperature to 260 °C, the temperature dependence of the specific heat capacity of the dry zeolite 13 X is crucial [142]. We thus use the equation provided by Simonot-Grange [142] (on page 225) for zeolite (zeo):

$$c_{\mathrm{p,zeo}} = 0.26\,\mathrm{J/g} + 8.6 \times 10^{-3}\,\mathrm{J/(g\,K)}\,(T_{\mathrm{zeo}} - 273.15\,\mathrm{K}) \tag{7.3}$$

to describe the specific heat capacity c_{ads} of the adsorbent (ads) in Equation (4.5).

7.1.2 Heat losses

With the thorough analysis of the heat losses of our adsorption-TES unit in Chapter 6, we are now able to apply experimentally validated heat-loss coefficients to the model. The heat-transfer coefficient from the casing to the environment $(UA)_{\mathrm{cas-env}}$(cf. Figure 7.2) is determined from the steady-state measurements for various temperatures in Section 6.1. The heat losses of the oil pipes $(UA)_{\mathrm{pipe-loss}}$ (cf. Figure 7.2) are also determined from steady-state measurements of the oil circulation system (cf. Appendix A.2). Thus, the heat-loss coefficients are available from independent measurements (cf. Section 6.1.1) and can be implemented into the model as a function of temperature according to the following equations with the constant heat-transfer coefficients $(UA)_{\mathrm{ce}}$ and $(UA)_{\mathrm{pl}}$ and the temperature-dependence factors f_{ce} and f_{pl}:

$$(UA)_{\mathrm{cas-env}} = (UA)_{\mathrm{ce}} + f_{\mathrm{ce}}\,(T_{\mathrm{cas}} - 273.15\,\mathrm{K}) \tag{7.4}$$

$$(UA)_{\mathrm{pipe-loss}} = (UA)_{\mathrm{pl}} + f_{\mathrm{pl}}\,(T_{\mathrm{oil}} - 273.15\,\mathrm{K}) \tag{7.5}$$

The heat-transfer coefficient from the adsorbent to the adsorber casing $(UA)_{\mathrm{ads-cas}}$ (cf. Figure 7.2) is also implemented as a function of the adsorbent temperature

$$(UA)_{\mathrm{ads-cas}} = (UA)_{\mathrm{ac}} + f_{\mathrm{ac}}\,(T_{\mathrm{ads}} - 273.15\,\mathrm{K}) \tag{7.6}$$

with a constant heat-transfer coefficient $(UA)_{\mathrm{ac}}$ and the temperature-dependent term with the factor f_{ac}. The constant values of $(UA)_{\mathrm{ac}}$ and f_{ac} are determined by model calibration in the following Section 7.2.

7.1.3 Evaporator model

The evaporator model is implemented with calibrated heat-transfer coefficients for vaporization $(UA)_{\mathrm{E,HX,evap}}$ and condensation $(UA)_{\mathrm{E,HX,cond}}$ (cf. Equation (4.12) and

Figure 7.2) to ensure that possible limitations of the heat transfer in the evaporator are described by the model. Moreover, the evaporator heat transfer depends on the fluid density $\varrho_{\mathrm{E,w}}$ in the evaporator

$$(UA)_{\mathrm{E,HX}} = (UA)_{\mathrm{E}} + f_{\mathrm{E}}\varrho_{\mathrm{E,w}} \tag{7.7}$$

with a constant heat-transfer coefficient $(UA)_{\mathrm{E}}$ and the constant density factor f_{E}. The fluid density $\varrho_{\mathrm{E,w}}$ is defined as mass of water per inner volume of the evaporator. Since the vapor density is negligible, $\varrho_{\mathrm{E,w}}$ is proportional to the water level in the evaporator. At high water level, the corrugated tubes on the three levels of the evaporator are mostly flooded with water. They fall dry with decreasing water level (cf. Section 3.2). The free surface of the evaporator tubes enhances condensation. Thus, the heat transfer during condensation $(UA)_{\mathrm{E,HX,cond}}$ increases with decreasing fluid density.

During vaporization, the opposite dependence on fluid density can generally be expected: the evaporator tubes have to be in contact with the adsorptive water to contribute to vaporization. Unfortunately, the corrugated pipe of the evaporator heat exchanger does not provide any capillary action. Thus, when the tubes fall dry at decreasing fluid densities, the heat transfer during vaporization $(UA)_{\mathrm{E,HX,evap}}$ decreases.

The general relation between heat transfer and fluid density is implemented in the model (cf. Equation (7.7)). However, the complex geometry of the evaporator (cf. Section 3.2) leads to inhomogeneous temperature and water distribution in the evaporator, leaving it infeasible to model the evaporator behavior physically correct with a lumped model of the VLE fluid and 1-D discretization of the heat exchanger. Thus, we cannot expect a perfect match between evaporator model and measurement. Still we achieve a more accurate description of the storage unit performance than in Chapter 4 without limitations of heat transfer in the evaporator model.

7.1.4 Connection of adsorber and evaporator

In contrast to the storage unit evaluated in Chapter 4, adsorber and evaporator are separated by a butterfly valve (cf. Figure 3.3). This valve between adsorber and evaporator limits the radiative heat transfer from the adsorber to the evaporator. Thus, heat transfer between adsorber and evaporator is neglected in the improved model, i.e. $\dot{Q}_{\mathrm{A-E}} = 0$ in the energy balances in Equations (4.5) and (4.10).

For the description of the mass transfer to the adsorbent, we still use the Glueckauf [120] approach with the effective mass transfer coefficient D_{eff} (cf. Equation (4.4)). No further mass transfer resistance is accounted for, if the valve is open, because the pressure drop induced by the valve is negligible. A closed valve, which is modeled as a flap (cf. Figure 7.2), interrupts the mass flow of the adsorptive water.

7.2 Calibration of the dynamic model

This section includes both the description of our calibration procedure and the results of calibrating our improved model to a measurement from Section 6.2.1. According to Trucano et al. [143], "calibration is to adjust a set of parameters associated with a computational science and engineering code so that the model agreement is maximized with respect to a set of experimental data." We accordingly choose a calibration procedure to determine the unknown parameters of the model of the adsorption thermal energy storage (TES) unit.

As experimental data, we choose to use measured values that are required in any case to evaluate the storage performance in terms of heat flow rates: temperatures of the in- and outlet flows of the heat exchangers and the respective fluid flow rates. Only one additional temperature measurement on the casing surface is required. Further temperature measurements are not necessary for the calibration procedure. Moreover, we use a simple pressure gauge which monitors the vacuum conditions inside the sealed casing.

The experimental setup (cf. Section 5.1) offers further temperature measurement positions, such as the temperature measurement on the adsorber surface (cf. Figure 5.1). However, this measurement position requires a thermocouple feed-through in the casing wall. Feed-throughs are prone to leakage and should thus be avoided, if possible. To prove that the feed-through can be avoided, we calibrate the model without the additional information on the adsorber surface temperature and still achieve a very good calibration result.

For model calibration, we use the measurement of a simple storage cycle with direct discharge (see Section 6.2.1): We charge the storage unit until the outlet flow of the adsorber reaches a desorption temperature of $T_{\text{des}} = 250\,°\text{C}$ and immediately switch to discharging. Discharging is stopped, when the outlet flow of the adsorber drops below a temperature of $T_{\text{ads}} = 70\,°\text{C}$. The calibration measurement is performed with an evaporator and condenser temperature of $T_{\text{evap}} = 40\,°\text{C}$.

The calibration approach is described in the following Section 7.2.1 with a stepwise determination of the model parameters for heat and mass transfer in the adsorber and evaporator. Section 7.2.2 contains the simulation of the calibration measurement, followed by a quantification of the simulation accuracy in Section 7.2.3.

7.2.1 Calibration approach

For a thorough calibration, we propose to determine the heat and mass transfer coefficients of the adsorption-TES model in 4 steps using steady-state and storage-cycle measurements from Chapter 6:

1. Heat-loss model + steady-state $\rightarrow (UA)_{\text{cas-env}}$
2. Evaporator model (cf. Figure 7.4) + storage-cycle $\rightarrow (UA)_{\text{E,HX}}$
3. Adsorber model (cf. Figure 7.5) + storage-cycle $\rightarrow (UA)_{\text{ads-cas}}$ and $(UA)_{\text{A,HX}}$
4. Storage-unit model (cf. Figure 7.2) + storage-cycle $\rightarrow D_{\text{eff}}$

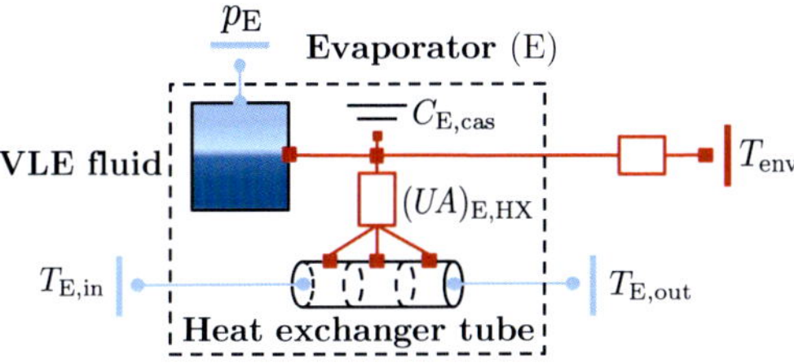

Figure 7.4: Evaporator model for separate calibration of $(UA)_{\text{E,HX}}$.

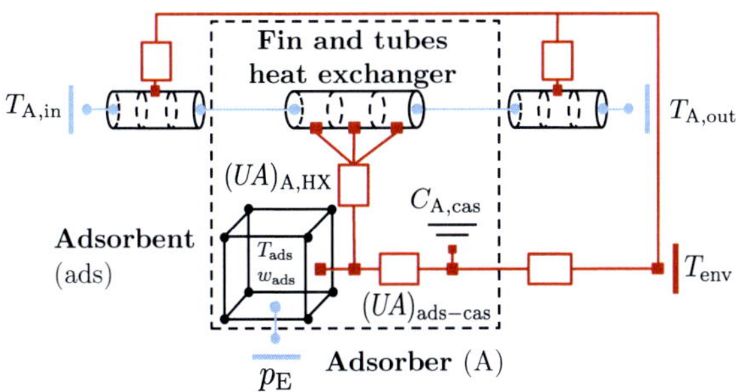

Figure 7.5: Adsorber model for separate calibration of $(UA)_{\text{ads-cas}}$ and $(UA)_{\text{A,HX}}$.

In step 1, we calculate the heat-transfer coefficient $(UA)_{\text{cas-env}}$ from steady-state measurement data. For the subsequent steps 2–4, we use measurement data from a storage cycle with charging and discharging. The transfer coefficients of the storage-unit model are calibrated separately for charging and discharging, since heat and mass transfer regimes may vary between ad- and desorption as well as between vaporization and condensation.

To be able to separately calibrate the evaporator and the adsorber model (steps 2 and 3), we need the state properties at the interface between the two models (cf. Figures 7.5 and 7.4). The measured value at this interface is the vapor pressure. However, the measured pressure at the interface between adsorber and evaporator has a large measurement uncertainty (cf. Appendix B.2). We further need the mass flow rate of the adsorptive water vapor $\dot{m}_{\text{ad}}$ at the interface between the two models. We can estimate this mass flow rate from an energy balance of the evaporator, but assumptions have to be taken. Due to these assumptions, and due to the measurement uncertainty of the pressure sensor, the results of the separate calibration of adsorber and evaporator are prone to uncertainty. Thus, the results of steps 2 and 3 are an estimate of the heat-transfer coefficients $(UA)_{\text{E,HX}}$ and $(UA)_{\text{A,HX}}$. Hence, these two coefficients will be adjusted in step 4, where we can use more-accurately measured values at the interfaces of the overall storage-unit model (cf. Figure 7.2).

In step 2, we estimate the magnitude of the density-dependent heat-transfer coefficient $(UA)_{\text{E,HX}}$ (cf. Equation (7.7)) from an energy balance of the evaporator heat exchanger.

In the steps 3 and 4, we calibrate the improved model (cf. Section 7.1) by simulating the calibration measurement with model input variables I from measurement data. We minimize the deviations between measured and simulated output variables Y to determine the remaining model parameters P of the adsorber model (cf. Figure 7.5) and of the storage-unit model (cf. Figure 7.2). Hence, we solve an optimization problem to calibrate the model. As input I and output variables Y, we choose directly measured values: temperatures, pressure and volume flow rates.

In step 3, we determine the temperature-dependent heat-transfer coefficient $(UA)_{\text{ads-cas}}$ (cf. Equation (7.6)) and estimate the magnitude of the constant heat-transfer coefficient $(UA)_{\text{A,HX}}$ from separate calibration of the adsorber model (cf. Figure 7.5).

Step 4 is the final calibration step: we use the storage-unit model (cf. Figure 7.2) to determine the last missing model parameter, i.e. the mass transfer coefficient D_{eff}. Simultaneously, we adjust the adsorber and evaporator heat-transfer coefficients $(UA)_{\text{E,HX}}$ and $(UA)_{\text{A,HX}}$ to compensate for the assumptions that we had to make to estimate these values. However, the estimates are crucial to our calibration procedure: The simultaneous calibration of heat and mass transfer cannot distinguish between the heat and mass transfer resistance in the adsorber and thus does not lead to distinct values, but to a ratio of the transfer coefficients $(UA)_{\mathrm{A,HX}}$ and D_{eff} [144]. In contrast, the separate calibration defines a physically reasonable range for the solution space of the heat transfer coefficient $(UA)_{\mathrm{A,HX}}$. Thus, the optimization of the storage-unit model finds the corresponding mass transfer coefficient in a physically consistent range.

The next paragraphs provide details of the calibration procedure. For the final calibration result, readers may proceed to Section 7.2.2.

Step 1: Heat-loss model calibration

The first calibration step is already described in Section 6.1.1, where we use steady-state measurements to determine the temperature-dependent heat-loss coefficient $(UA)_{\mathrm{cas-env}}$: Heat losses of the adsorber are calculated from measurements with constant temperatures in the adsorber and for different vapor pressures. From these steady-state measurements, the heat transfer outside of the adsorber casing is determined. Figure 6.2 shows the resulting fit for $(UA)_{\mathrm{cas-env}}$ as a linear function of the casing temperature (cf. Equation (7.4)). Numerical results are listed in Table 7.3.

Heat losses of the oil pipes are calculated from measurements at constant temperatures in the oil circulation system, while the adsorber is bypassed. Results for the loss coefficient of the oil pipes $(UA)_{\mathrm{pipe-loss}}$ (cf. Equation (7.5)) to the environment can be found in Appendix A.2.

For the following calibration steps, we use the measurement data from the calibration measurement, i.e. from a direct-discharge measurement with $T_{\mathrm{des}} = 250\,°\mathrm{C}$, $T_{\mathrm{ads}} = 70\,°\mathrm{C}$ and $T_{\mathrm{cond}} = T_{\mathrm{evap}} = 40\,°\mathrm{C}$ (cf. Section 6.2.1).

Estimating the mass flow rate of adsorptive water

To allow separate calibration of the evaporator and the adsorber model (steps 2 and 3), we need the mass flow rate of adsorptive water vapor $\dot{m}_{\mathrm{ad}}$. To calculate $\dot{m}_{\mathrm{ad}}$, we use measured values from the calibration measurement: the mass flow rate $\dot{m}_{\mathrm{E,HX}}$, the in- and outlet temperature of the evaporator heat exchanger $T_{\mathrm{E,in}}$ and $T_{\mathrm{E,out}}$, and the pressure p_{E}, which equals the pressure at the interface between adsorber and evaporator model (cf. Figures 7.5 and 7.4). However, it has to be noted that the pressure sensor has a large measurement uncertainty (cf. Appendix B.2) compared to the uncertainty of the temperatures and the fluid flow rates.

The mass flow rate of adsorptive water vapor $\dot{m}_{\mathrm{ad}}$ is calculated with the energy balance of the evaporator:

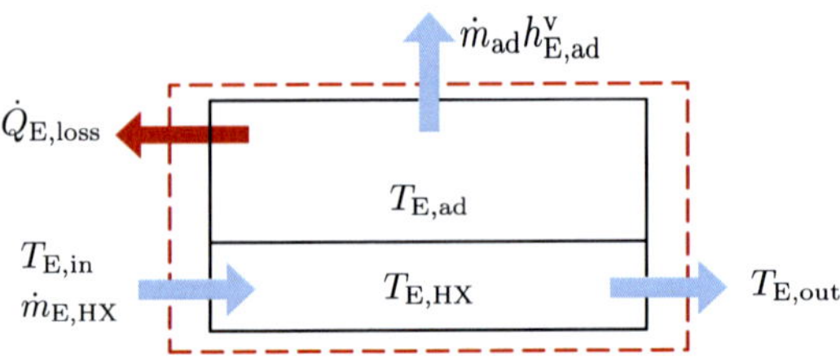

$$\frac{\mathrm{d}U_{\mathrm{E,HX}}}{\mathrm{d}t} + \left(c_{\mathrm{E}}\, m_{\mathrm{E,ave}}\right) \frac{\mathrm{d}T_{\mathrm{E,ad}}}{\mathrm{d}t} + u^{\mathrm{l}}_{\mathrm{E,ad}} \frac{\mathrm{d}m_{\mathrm{E,ad}}}{\mathrm{d}t}$$
$$= \dot{m}_{\mathrm{E,HX}} \left(h_{\mathrm{E,in}} - h_{\mathrm{E,out}}\right) - \dot{m}_{\mathrm{ad}}\, h^{\mathrm{v}}_{\mathrm{E,ad}} - \dot{Q}_{\mathrm{E,loss}}\,, \tag{7.8}$$

with the internal energy of the evaporator heat exchanger $U_{\mathrm{E,HX}}$, the heat capacity c_{E} and the average mass $m_{\mathrm{E,ave}}$ of adsorptive water, evaporator casing and steel trays (cf. Figure 3.2, b):

$$\left(c_{\mathrm{E}}\, m_{\mathrm{E,ave}}\right) = \left(c_{\mathrm{E,ad}}\, m_{\mathrm{E,ad,ave}} + c_{\mathrm{steel}}\, m_{\mathrm{E}}\right), \tag{7.9}$$

the temperature $T_{\mathrm{E,ad}}$, the mass $m_{\mathrm{E,ad}}$ and internal energy $u^{\mathrm{l}}_{\mathrm{E,ad}}$ of the adsorptive water in the evaporator, the mass flow rate and enthalpy change $\dot{m}_{\mathrm{E,HX}} \left(h_{\mathrm{E,in}} - h_{\mathrm{E,out}}\right)$ of the water flowing through the evaporator heat exchanger, the mass flow rate $\dot{m}_{\mathrm{ad}}$ and enthalpy $h^{\mathrm{v}}_{\mathrm{E,ad}}$ of adsorptive water vapor leaving the evaporator and the heat loss of the evaporator $\dot{Q}_{\mathrm{E,loss}}$:

$$\dot{Q}_{\mathrm{E,loss}} = (UA)_{\mathrm{cas-env}} \left(T_{\mathrm{E,ad}} - T_{\mathrm{env}}\right). \tag{7.10}$$

We assume that the adsorptive water inside the evaporator has a homogeneous temperature $T_{\mathrm{E,ad}}$, which we assume to be equal to the temperature of the evaporator casing and the steel trays (cf. Figure 7.4). Since the measured pressure p_{E} equals the saturation pressure of the adsorptive water in the evaporator, the temperature of the adsorptive water is $T_{\mathrm{E,ad}} = T_{\mathrm{s}} \left(p_{\mathrm{E}}\right)$ (cf. Equation (4.2)).

Further, we assume that the water in the heat exchanger behaves as an ideal fluid and that the pressure drop inside the heat exchanger tube is negligible, leading to:

$$\dot{m}_{\mathrm{E,HX}} \left(h_{\mathrm{E,in}} - h_{\mathrm{E,out}}\right) = \dot{m}_{\mathrm{E,HX}}\, c_{\mathrm{p,w}} \left(T_{\mathrm{E,in}} - T_{\mathrm{E,out}}\right) \tag{7.11}$$

with the mass flow rate $\dot{m}_{\mathrm{E,HX}}$, specific heat capacity $c_{\mathrm{p,w}}$, the inlet $T_{\mathrm{E,in}}$ and outlet temperature $T_{\mathrm{E,out}}$ of the water in the evaporator heat exchanger.

We also assume that the mass and heat capacity inside the evaporator heat exchanger tube are constant and we thus approximate the change of internal energy by

$$\frac{\mathrm{d}U_{\mathrm{E,HX}}}{\mathrm{d}t} = (m\,c_{\mathrm{p}})_{\mathrm{E,HX}} \left(\frac{\mathrm{d}T_{\mathrm{E,HX}}}{\mathrm{d}t} \right) \tag{7.12}$$

assuming that the temperature gradient of all parts of the evaporator heat exchanger is equal to the gradient of an average temperature $T_{\mathrm{E,HX}}$, which we calculate by $T_{\mathrm{E,HX}} = 0.5\,(T_{\mathrm{E,in}} + T_{\mathrm{E,out}})$. The term $(m\,c_{\mathrm{p}})_{\mathrm{E,HX}}$ is the mass and specific heat capacity of the evaporator heat exchanger, including the tube wall and water.

Moreover, if we assume that $u^{\mathrm{l}}_{\mathrm{E,ad}} \approx h^{\mathrm{l}}_{\mathrm{E,ad}}$, we gain the enthalpy of vaporization $\Delta h_{\mathrm{v}} = h^{\mathrm{v}}_{\mathrm{E,ad}} - h^{\mathrm{l}}_{\mathrm{E,ad}}$. As the decrease of adsorptive water mass in the evaporator is equal to the mass flow rate of adsorptive water vapor (cf. Equation (7.18)), we finally get:

$$\dot{m}_{\mathrm{ad}} = \frac{1}{\Delta h_{\mathrm{v}}} \left(\dot{m}_{\mathrm{E,HX}}\,(h_{\mathrm{E,in}} - h_{\mathrm{E,out}}) - (c_{\mathrm{E}}\,m_{\mathrm{E,ave}}) \frac{\mathrm{d}T_{\mathrm{E,ad}}}{\mathrm{d}t} - \frac{\mathrm{d}U_{\mathrm{E,HX}}}{\mathrm{d}t} - \dot{Q}_{\mathrm{E,loss}} \right), \tag{7.13}$$

with the results from Equations (7.9)–(7.12). The mass flow rate $\dot{m}_{\mathrm{ad}}$ can now be used to model adsorber and evaporator separately in the next two calibration steps 2 and 3. However, the various assumptions leading to Equation (7.13) are prone to uncertainty, in particular the assumption of homogeneous temperatures. Thus, the results from the following calibration steps 2 and 3 should be adjusted in step 4.

Step 2: Evaporator model calibration

For the estimation of the evaporator heat-transfer coefficient $(UA)_{\mathrm{E,HX}}$, we again use the measured in- and outlet temperature of the evaporator and the pressure. We derive the heat-transfer coefficient $(UA)_{\mathrm{E,HX}}$ from an energy balance around the evaporator heat exchanger tube:

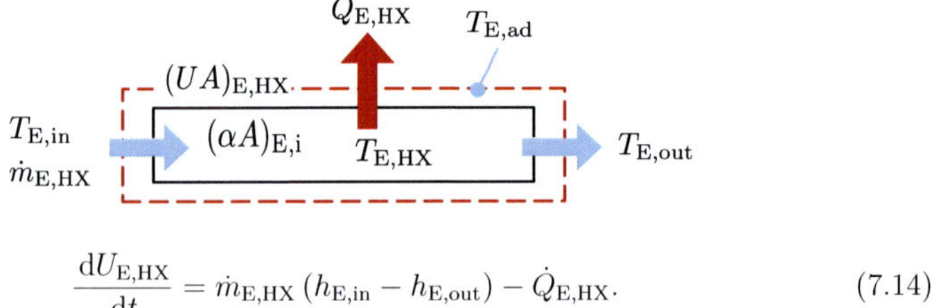

$$\frac{\mathrm{d}U_{\mathrm{E,HX}}}{\mathrm{d}t} = \dot{m}_{\mathrm{E,HX}} \left(h_{\mathrm{E,in}} - h_{\mathrm{E,out}}\right) - \dot{Q}_{\mathrm{E,HX}}. \tag{7.14}$$

The heat flow rate across the evaporator heat exchanger wall $\dot{Q}_{\mathrm{E,HX}}$ can be calculated by

$$\dot{Q}_{\mathrm{E,HX}} = (UA)_{\mathrm{overall}} \Delta T_{\mathrm{ln}} = (UA)_{\mathrm{overall}} \frac{T_{\mathrm{E,in}} - T_{\mathrm{E,out}}}{\ln\left(T_{\mathrm{E,in}} - T_{\mathrm{E,ad}}\right) - \ln\left(T_{\mathrm{E,out}} - T_{\mathrm{E,ad}}\right)}. \tag{7.15}$$

In Equation (7.15), ΔT_{ln} is the logarithmic temperature difference, which assumes a homogeneous temperature on the outer surface of the heat exchanger tube [76]. This temperature is equal to the temperature of adsorptive water $T_{\mathrm{E,ad}}$ (cf. Equation (4.2)).

The overall heat-transfer coefficient $(UA)_{\mathrm{overall}}$ from the evaporator heat exchanger to the adsorptive water combines the heat-transfer coefficient of the evaporator heat exchanger on the outer tube wall with the inner heat-transfer coefficient $\alpha_{\mathrm{E,i}}A$ [129]:

$$(UA)_{\mathrm{overall}} = \left(\frac{1}{(UA)_{\mathrm{E,HX}}} + \frac{1}{\alpha_{\mathrm{E,i}}A}\right)^{-1}. \tag{7.16}$$

We hereby neglect the heat-transfer resistance of the tube wall, because it is very low compared to the heat-transfer resistances in the fluid and in the adsorbent. The heat-transfer coefficient $\alpha_{\mathrm{E,i}}$ is calculated with Nusselt correlations for tubular fluid flow [122].

Based on the mentioned assumptions, Equations (7.14)–(7.16) allow the quantification of $(UA)_{\mathrm{E,HX}}$ by:

$$(UA)_{\mathrm{E,HX}} = \left(\left[\frac{\dot{m}_{\mathrm{E,HX}} \left(h_{\mathrm{E,in}} - h_{\mathrm{E,out}}\right) - \frac{\mathrm{d}U_{\mathrm{E,HX}}}{\mathrm{d}t}}{\Delta T_{\mathrm{ln}}}\right]^{-1} - \frac{1}{\alpha_{\mathrm{E,i}}A}\right)^{-1} \tag{7.17}$$

together with the results from Equations (7.11) and (7.12).

As discussed in Section 7.1.3, we expect the heat-transfer coefficient of the evaporator $(UA)_{\mathrm{E,HX}}$ to depend on the adsorptive water level in the evaporator, i.e. on the fluid density (cf. Equation (7.7)). To determine this dependence, we need the fluid density as a function of time $\varrho_{\mathrm{E,w}}(t)$. With the constant volume of the evaporator V_{evap}, the change in density is proportional to the mass flow rate of adsorptive water vapor $\dot{m}_{\mathrm{ad}}$ that leaves the evaporator:

$$\frac{\mathrm{d}\varrho_{\mathrm{E,w}}}{\mathrm{d}t} = -\frac{1}{V_{\mathrm{evap}}}\frac{\mathrm{d}m_{\mathrm{E,ad}}}{\mathrm{d}t} = -\frac{\dot{m}_{\mathrm{ad}}}{V_{\mathrm{evap}}}. \tag{7.18}$$

The mass flow rate $\dot{m}_{\mathrm{ad}}$ can now be used to determine the fluid density in the evaporator for each time step (cf. Equation (7.18)).

Figure 7.6 shows the results of the calibration of the evaporator model, in particular the results of Equation (7.17) with the heat-transfer coefficient in the evaporator as a function of the adsorptive water fluid density $(UA)_{\mathrm{E,HX}} = f(\varrho_{\mathrm{E,w}})$ (cf. Equation (7.7)). As expected (cf. Section 7.1.3), the heat-transfer coefficient follows a linear falling trend

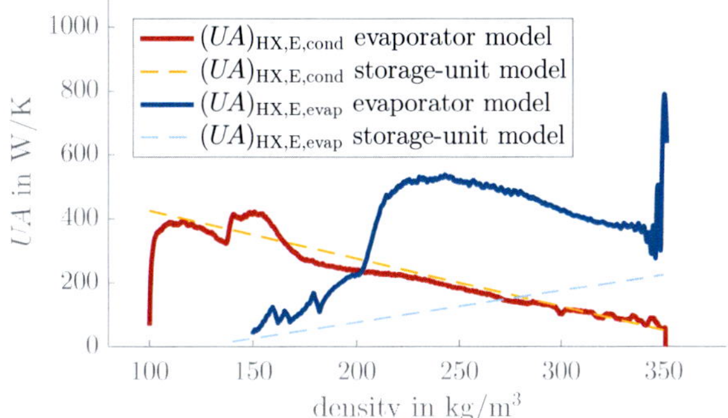

Figure 7.6: Heat transfer coefficient for condensation and vaporization as a function of fluid density in the evaporator. Solid lines result from Equation (7.17), dashed lines show the heat-transfer coefficients implemented in the model after calibration of the storage-unit model.

during condensation. The heat transfer during vaporization is obviously limited at low fluid densities, as expected. However, for rising fluid densities, the heat-transfer coefficient $(UA)_{\mathrm{E,HX,evap}}$ (evaporator model, Equation (7.17)) does not monotonically increase, but even decreases. However, we decided not to model this inverse trend, because the physical reasons are not clear and large systematic uncertainties have to be expected for the input variables to Equation (7.17) due to the manifold assumptions.

The adsorptive water is not distributed evenly to all three evaporator levels (cf. Section 3.2) and we cannot assure that the water flow is distributed evenly through all three parallel heat exchanger tubes. Thus, the assumption of a homogeneous tube surface temperature is inaccurate. In particular during high flow rates of adsorptive water at the beginning of vaporization, we expect the water surface to cool down more quickly than the evaporator sump, leading to large temperature gradients. However, the lumped model does not allow for describing these effects. Moreover, temperature of the water is calculated from the pressure measurement, which adds large measurement uncertainty (cf. Appendix B.2).

The calculated heat-transfer coefficient of the evaporator heat exchanger $(UA)_{\mathrm{E,HX}}$ provides an estimate of the correct physical range of the parameter value. It is eventually adjusted later while calibrating the storage-unit model. During calibration of the storage-unit model, we use measurement data with high accuracy as input variables leading to an accurate calibration result. Hence, Figure 7.6 also contains the curve for the coefficient $(UA)_{\mathrm{E,HX}}$ which is implemented in the final model. The coefficient of heat transfer in the evaporator $(UA)_{\mathrm{E,HX}}$ is described with a linear dependence on the water fluid density (cf. Equation (7.7)).

Optimization problem

For the subsequent steps 3 and 4, we solve optimization problems to calibrate the remaining model parameters (P). To solve the optimization problems, we analyze the entire solution space with a full factorial design to find optimal model parameters (P). Our dynamic model, containing the differential states X and the algebraic states Z, is a nonlinear system of differential algebraic equations (DAE). The DAE system can generally be written as

$$\begin{aligned} \frac{\mathrm{d}}{\mathrm{d}t} X(t) &= \mathcal{F}(X(t),\, I(t),\, Z(t),\, P) \\ 0 &= f(Z(t), X(t)) \end{aligned}$$

As the objective function for the optimization, we define the root mean square deviation $RMSD$ of the measured and simulated output variable Y normalized with its measurement uncertainty u_Y (cf. Appendix B):

$$RMSD_{\mathrm{norm}}(Y) = \sqrt{\frac{\int \left[\left(Y_{\mathbf{meas}}(t) - Y_{\mathbf{sim}}(t) \right) / u_Y(t) \right]^2 \, \mathrm{d}t}{\Delta t}}). \tag{7.19}$$

By referring to the uncertainty of the measured values, we create a dimensionless objective function that takes into account the data quality.

If we use more than one output variable Y for calibration, we calculate the arithmetical mean of the normalized root mean square deviations of the output variables

$$\overline{RMSD}_{\text{norm}} = \frac{1}{N}\sum_{i=1}^{N} RMSD_{\text{norm}}(Y_i)$$

with N being the number of output variables used for calibration. Further weighting is not necessary, since the root mean square deviations $RMSD_{\text{norm}}$ are already normalized with the corresponding measurement uncertainty.

The optimization problem finally reads:

$$\min_{P} \overline{RMSD}_{\text{norm}}(X(t),\, I(t),\, Z(t),\, P) \tag{7.20}$$

$$\begin{aligned} \text{s.t.}\ \ & \dot{X}(t) = \mathcal{F}(X(t),\, I(t),\, Z(t),\, P) \\ & X(t=0) = X_0, \\ & P_{\min} \leq P \leq P_{\max} \end{aligned} \tag{7.21}$$

which we solve by full factorial design [145]: A parameter sweep is performed with reasonable values of the missing model parameters P. GenOpt® [146] is used to raster the results of the objective function $RMSD_{\text{norm}}(Y)$ for all possible combinations of the parameters P. We thus avoid local minima of our objective function. Moreover, the full factorial design allows for a sensitivity analysis of the results.

Step 3: Adsorber model calibration

Calibration of the adsorber model is performed to determine the model parameters P_{adsorber} for heat transfer in the adsorber: from the adsorber heat exchanger to the adsorbent $(UA)_{\text{A,HX}}$ and from the adsorbent to the casing $(UA)_{\text{ads-cas}}$. For this purpose, the adsorber model (cf. Figure 7.5) is simulated separately as follows: the input variables I_{adsorber} to the simulation of the adsorber model are the mass flow rate $\dot{m}_{\text{oil}}$ and the input temperature $T_{\text{A,in}}$ of the oil, the ambient temperature T_{env} and the enthalpy h^{v}_{ad} and mass flow rate of the adsorptive water $\dot{m}_{\text{ad}}$ (cf. Equation (7.13)).

During condensation, the enthalpy of the water vapor h^{v}_{ad} which leaves the adsorber is assumed to have the same temperature as the adsorber T_{ads}. Incoming water vapor during vaporization is modeled assuming that the temperature of water leaving the evaporator is equal to the saturation temperature.

The output variables Y_{adsorber} for calibration of the adsorber model are the output temperature and the casing temperature. Finally, we have the differential states of the adsorber model X_{adsorber} and the algebraic states Z_{adsorber}, listed in Table 7.1.

Table 7.1: Variables for the optimization problem to calibrate the adsorber model.

model parameter P_{adsorber}	$(UA)_{\text{A,HX}}$, $(UA)_{\text{ads-cas}} = f((UA)_{\text{ac}}, f_{\text{ac}})$
input variables I_{adsorber}	$\dot{m}_{\text{oil}}$, $T_{\text{A,in}}$, T_{env}, $\dot{m}_{\text{ad}}$
differential states X_{adsorber}	T_{A}, w_{A}, $T_{\text{A,HX,i}}$, $T_{\text{A,cas}}$
algebraic states Z_{adsorber}	properties of all media in the adsorber
output variables Y_{adsorber}	$T_{\text{A,out}}$, $T_{\text{A,cas}}$

The heat-transfer coefficient of the adsorber heat exchanger $(UA)_{\text{A,HX}}$ is assumed to be constant and the heat-transfer coefficient from the adsorbent to the casing $(UA)_{\text{ads-cas}}$ is temperature dependent with the constant $(UA)_{\text{ac}}$ and temperature-dependence factor f_{ac} (cf. Equation (7.6)). The parameter sweep is started with a broad range of possible heat-transfer coefficients varying from 1 to 500 W/K and temperature dependence-factor from 0 to 1 W/K^2. This broad range of values ensures that the minimum of the objective function $\overline{RMSD}_{\text{norm}}(T_{\text{A,out}}, T_{\text{A,cas}})$ is found in the optimization. Results of the calibration of the adsorber model are shown in Figure 7.7 as contour plot with a variation of the heat-

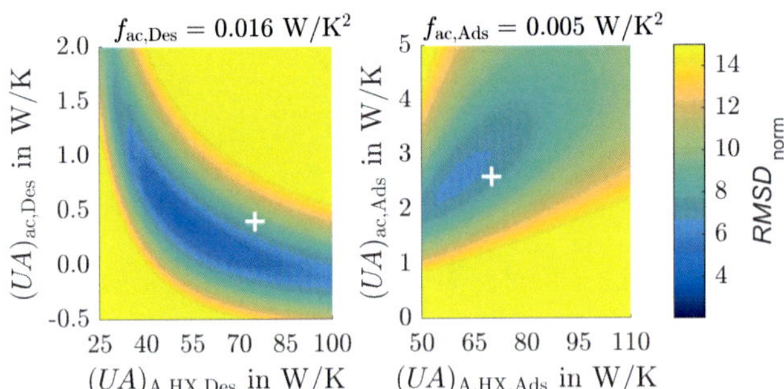

Figure 7.7: Results of adsorber model calibration for desorption (left) and adsorption (right): Heat transfer of adsorber heat exchanger to adsorbent $(UA)_{\text{A,HX}}$ and adsorbent to casing $(UA)_{\text{ads-cas}}$ according to Equation (7.6) with the constant $(UA)_{\text{ac}}$ and factor f_{ac}. Contour plots show the arithmetical mean of the normalized root mean square deviations of measured and simulated adsorber output temperature and casing temperature $\overline{RMSD}_{\text{norm}}(T_{\text{A,out}}, T_{\text{A,cas}})$; white crosses result from calibration of the storage-unit model.

transfer coefficients $(UA)_{\text{A,HX}}$ and $(UA)_{\text{ac}}$ for the best fit of the temperature-dependence factor f_{ac}. The results of the heat-transfer coefficients $(UA)_{ads-cas}$ (cf. Equation (7.6)) are also shown in Figure 7.8 as a function of the adsorber temperature. The results for de- and adsorption are in the same order of magnitude, which is physically consistent.

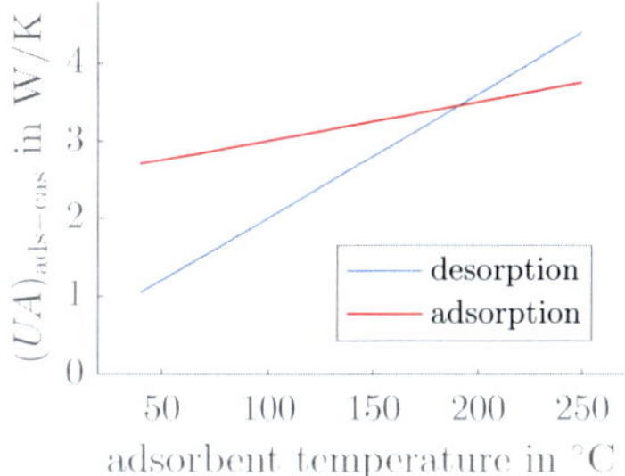

Figure 7.8: Heat transfer coefficient from adsorber to casing $(UA)_{ads-cas}$ as function of the adsorber temperature T_{ads} for de- and adsorption: results of the calibration of the storage-unit model.

The calibration of the adsorber model relies on the estimation of the mass flow rate of the adsorptive water $\dot{m}_{ad}$ (cf. Equation (7.13)). Since the calculation of $\dot{m}_{ad}$ is prone to uncertainties, we adjust the heat-transfer coefficients $(UA)_{A,HX}$ again in the next calibration step. Nevertheless, we do not recalibrate $(UA)_{ac}$, because the driving temperature difference for the heat flow from the adsorber to the casing is sufficiently large (cf. Figure 6.1) and does not vary significantly with $\dot{m}_{ad}$.

Step 4: Storage-unit model calibration

Finally, the last calibration step combines adsorber and evaporator to the storage-unit model to determine the coefficient D_{eff} for mass transfer in the adsorber. The input variables $I_{storage-unit}$ to the simulation with the storage-unit model are the mass flow rates and the input temperatures of oil and water as well as the ambient temperature, cf. Table 7.2. The output variables $Y_{storage-unit}$ for calibration are the adsorber output temperature and casing temperature, the evaporator output temperature and the pressure inside the evaporator. These variables are listed in Table 7.2 together with the differential $X_{storage-unit}$ and algebraic states $Z_{storage-unit}$ of the storage-unit model.

The model parameter of the calibration of the storage-unit model are the mass transfer coefficient D_{eff} and the heat-transfer coefficients in the adsorber $(UA)_{A,HX}$ and the

Table 7.2: Variables for the optimization problem to calibrate the storage-unit model

model parameter $P_{\text{storage-unit}}$	D_{eff}, $(UA)_{\text{A,HX}}$, $(UA)_{\text{E,HX}} = f((UA)_{\text{E}}, f_{\text{E}})$
input variables $I_{\text{storage-unit}}$	$\dot{m}_{\text{oil}}$, $T_{\text{A,in}}$, T_{env}, $\dot{m}_{\text{E,HX}}$, $T_{\text{E,in}}$
differential states $X_{\text{storage-unit}}$	ϱ_{E}, T_{E}, T_{A}, w_{A}, $T_{\text{A,HX,i}}$ $T_{\text{E,HX,i}}$, T_{Valve}, $T_{\text{A,Cas}}$
algebraic states $Z_{\text{storage-unit}}$	properties of all media in adsorber and evaporator
output variables $Y_{\text{storage-unit}}$	$T_{\text{A,out}}$, $T_{\text{A,cas}}$, $T_{\text{E,out}}$, p_{A}

evaporator $(UA)_{\text{E,HX}}$ (cf. Equation (7.7)). We adjust the heat-transfer coefficients once again (compare P_{adsorber} and $P_{\text{storage-unit}}$ in Table 7.1 and 7.2), since the input variables $I_{\text{storage-unit}}$ in this final calibration step are all measured directly and thus more accurate. In contrast, the rather inaccurate vapor flow rate of adsorptive water $\dot{m}_{\text{ad}}$ (cf. Equation (7.13)) was used to split the model into evaporator and adsorber to separately determine the heat-transfer coefficients.

For the full factorial design, the parameter values are varied in the following range: the constants of the heat transfer coefficient $(UA)_{\text{E,evap/cond}}$ in the evaporator (cf. Equation (7.7)) are varied from 0 to 1200 W/K and the density-dependence factors $f_{\text{E,evap/cond}}$ are varied between -5 and 5 W m^3/(K kg). The heat-transfer coefficients in the adsorber $(UA)_{\text{A,HX,des/ads}}$ are varied from 30 to 100 W/K. The mass transfer coefficients $D_{\text{eff,des/ads}}$ are varied from 1×10^{-11} to 1×10^{-7} m/s^2.

In the calibration of the storage-unit model, the de- and adsorption phase are calibrated simultaneously, because separate calibration of the desorption parameters leads to bad initial conditions for adsorption. Hence, the optimization problem has eight variables $P_{\text{storage-unit}}$ as degrees of freedom. The eight-dimensional result cannot be depicted graphically, but is given in Table 7.3, where the final calibration results are listed. These coefficients are implemented in the model.

Table 7.3: Model parameters from calibration.

Coefficient	Desorption	Adsorption	Unit
$(UA)_{\text{cas-env}}$	$0.6789 + 0.0036\ T_{\text{cas}}$		W/K
$(UA)_{\text{E,HX}}$	$575 - 1.5\ \varrho_{\text{E,ad}}$	$-125 + 1\ \varrho_{\text{E,ad}}$	W/K
$(UA)_{\text{A,HX}}$	75	70	W/K
$(UA)_{\text{ads-cas}}$	$0.4 + 0.016\ T_{\text{ads}}$	$2.6 + 0.005\ T_{\text{ads}}$	W/K
D_{eff}	1e-7	1e-7	m^2/s^2

Mass transfer coefficients larger than 1×10^{-7} m/s^2 do not alter the simulation results, as mass transfer does not limit the process anymore. Thus, our result of the mass transfer coefficients $D_{\text{eff}} = 1 \times 10^{-7}\,\text{m/s}^2$ during desorption and adsorption confirms what is consensus in literature: the main limitation is the heat transfer [89], especially for loose grain configurations [12].

The final values of the heat-transfer coefficients are close to the results from the separate adsorber and evaporator model calibration. Figure 7.7 shows that the final values for $(UA)_{\text{A,HX}}$ are only slightly altered compared to the calibration of the adsorber model. For the charging period, i.e. during condensation, the heat-transfer coefficient of the evaporator $(UA)_{\text{E,HX,cond}}$ is very close to the results from separate calibration (cf. red lines in Figure 7.6). The corresponding heat-transfer coefficient during discharging is slightly adjusted. The final result for the heat-transfer coefficient $(UA)_{\text{E,HX,evap}}$ (blue dashed line in Figure 7.6) shows that the heat transfer during vaporization limits the process at low fluid densities and increases with fluid density, just as we expected.

7.2.2 Simulation results of the calibration measurement

Figure 7.9 shows the simulation results for the calibration measurement. Deviations

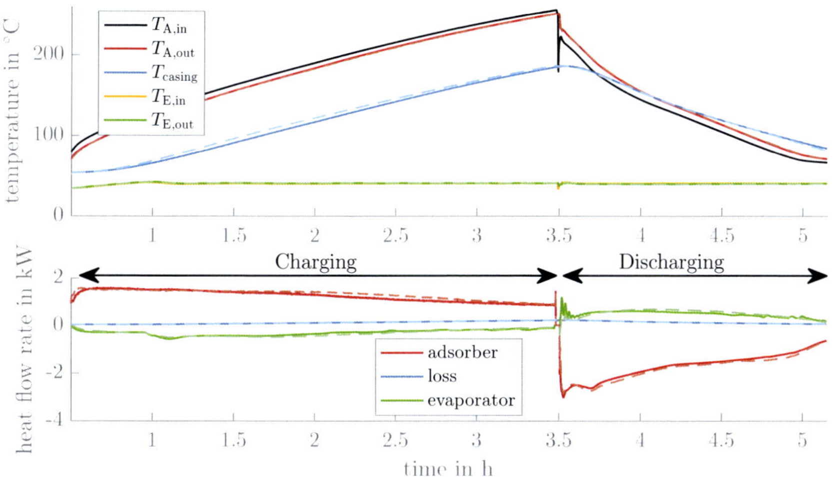

Figure 7.9: Calibration measurement with direct discharge, $T_{\text{des}} = 250\,°\text{C}$ and $T_{\text{evap}} = 40\,°\text{C}$. Top: temperatures of adsorber and evaporator. Bottom: heat flow rate of adsorber, evaporator and losses to the environment. Dashed lines show the corresponding simulation results.

between measured (full lines in Figure 7.9) and simulated temperatures (dashed lines in Figure 7.9) are hardly visible. Thus, the calibrated model provides a very good agreement between measurement and simulation. Based on these temperatures and the measured flow rates of oil and water, the heat flow rates can be calculated: $\dot{Q}_{\mathrm{E,HX}}$ from Equation (7.14), $\dot{Q}_{\mathrm{A,HX}}$ according to Equation (6.11) and the heat-loss rate according to Equation (6.13) (cf. Section 6.2).

Figure 7.9 (bottom) shows the measured and simulated heat flow rates to the adsorber and evaporator as well as the heat-loss rate calculated with Equation (6.13). The measurement data is well reproduced by the calibrated model despite the simplified geometry and the few calibrated heat and mass transfer coefficients. We have also checked to describe the transfer coefficients inside the adsorber $(UA)_{\mathrm{A,HX}}$ and D_{eff} as depending on water content w_{ads} or temperature T_{ads} of the adsorbent, but the resulting simulation does not give better results than the constant values that we finally used (cf. Table 7.3).

Figure 7.9 reveals some small deviations between simulation and experiment. These deviations are mainly caused by the simplified geometry and by the representation of the equilibrium data based on only few measurements of the adsorber equilibrium loading. An error in the equilibrium data, i.e. the adsorption characteristics, may influence the calibration of heat and mass transfer inside the adsorber and evaporator: the model for the adsorption process could tend to underestimate the adsorption potential and thus the heat release at low water contents in the adsorbent. The calibration can partly compensate such a mismatch in the description of the adsorption process by adjusting the heat and mass transfer: if heat release by adsorption is underestimated at the beginning of discharging, the heat-transfer coefficients may be overestimated in the calibration procedure. As a result, the heat flow rate (cf. Figure 7.9, bottom) is slightly overestimated by the model as soon as the adsorption process slows down.

The simplified geometry description of the lumped model approach affects the heat transfer performance of the adsorber unit. In the model, all thermal capacities in the adsorber, i.e. the complete zeolite and all the metal parts are heated and cooled at the same time. In reality, e.g. during adsorption, first the outer part of the zeolite is exposed to water vapor. Thus, the outer shell heats up first, so the adsorber faces a high temperature in the outer zeolite beads at the beginning of adsorption. As the metal parts of the heat exchanger have a good thermal conductance, the temperature rise can quickly be transferred to the oil tube. The measured heat flow rate $\dot{Q}_{\mathrm{A,HX}}$ has a peak then. The same effect applies for the desorption at the beginning of the measurement where a sharp change in oil temperature and good thermal transfer in the metal parts of the heat exchanger lead to high heat flow rates (cf. Figure 7.9, bottom). These high heat flow rates are not exactly reproduced by the model as it does not distinguish between metal and zeolite, but assumes a homogeneous material. As a result of calibration, by minimizing the deviations between measurement and simulation, we tend to overestimate the heat-transfer coefficient of the adsorber heat exchanger in order to compensate for the underestimated heat flow rate at the beginning of de- and adsorption. Figure 7.9 (bottom) reveals that this underestimation of the heat flow rate lasts for around three minutes and then the curve of the simulated heat flow rates crosses the measured curve. Subsequently, the model slightly overestimates the heat flow rates at the end of each de- and adsorption.

The evaporator heat flow rate $\dot{Q}_{\mathrm{E,HX}}$ follows the same trend as the adsorber: heat flow rates are slightly underestimated at the beginning of vaporization and condensation and slightly overestimated later. This trend is determined by the adsorber, because desorption and adsorption influence the vapor pressure and cause the water vapor to condense and vaporize.

Despite small deviations, the model correctly describes the dynamics of the storage process (cf. Figure 7.9, bottom): the simulated heat flow rates closely follow the trend of the measured values. The heat losses are very accurately described by the model, since the heat-loss coefficients outside of the adsorber are calibrated independent of processes inside the adsorber, cf. Section 6.1. The accuracy of the heat losses is mainly determined by the casing temperature T_{cas}: our model correctly describes the trend of T_{cas}, which is used to calculate the heat-loss rate (cf. Equation (6.13)). Thus, the heat-loss rate is influenced by the heat transfer to the casing, which is described with the coefficient $(UA)_{\mathrm{ads-cas}}$.

The calibration results of $(UA)_{\mathrm{ads-cas}}$ may be influenced by the lumped model approach. In particular, the assumption of a homogeneous temperature of the surface of the adsorber sheet and casing is not accurate, as measurements with redundant thermometers (cf. Section 5.1) showed gradients of more than 10 K over the casing surface. However, despite these gradients of the casing temperature, the heat flow rate from the adsorber to the casing is accurately predicted, because the driving temperature difference between adsorber surface and casing (cf. Figure 6.1) is sufficiently large. Thus, the visible deviation between measured and simulated casing temperature during desorption in Figure 7.9 (top) is not critical. During adsorption, the simulated casing temperature and thus the heat losses are precisely described by the model.

Altogether, the model describes the temperatures and heat flow rates of the calibration measurement very accurately. The small visual deviations are quantified and compared to the measurement uncertainty in the following Section 7.2.3.

7.2.3 Simulation accuracy measures

Comparing the simulation result visually to the measured result gives an idea of the model accuracy. Though, it remains difficult to compare to other storage models, if only temperature curves are shown (cf. Section 2.5.4). Consequently, we use statistical measures to quantify the accuracy of the model and at the same time make the accuracy comparable. Since the performance of a storage unit is defined by heat flow rates and transferred energies, we quantify the simulation accuracy with two statistical measures that describe (1) the accuracy of the simulated heat flow rates and (2) the accuracy of the simulated transferred energies.

The accuracy of the heat flow rates, i.e. the dynamics of the system, is given by the coefficient of variation CV of the heat flow rates. The CV was already introduced by Lanzerath [93]. As defined in Equation (2.5), the CV is based on the root mean square deviation $RMSD$ divided by the average heat flow rate.

The accuracy of the thermal energy, e.g. to describe the efficiency or the energy storage density, is given by the relative deviation of the transferred thermal energy ΔE_{rel}, which we define as

$$\Delta E_{\mathrm{rel}} = \frac{\int \dot{Q}_{\mathrm{meas}}\,\mathrm{d}t - \int \dot{Q}_{\mathrm{sim}}\,\mathrm{d}t}{\int \dot{Q}_{\mathrm{meas}}\,\mathrm{d}t}. \tag{7.22}$$

The coefficient of variation CV of the relevant heat flow rates for the calibration measurement is shown in Figure 7.10a and the relative deviation of the thermal energy ΔE_{rel} for the calibration measurement is plotted in Figure 7.10b. Furthermore, the relative measurement uncertainty (cf. Appendix B) is plotted and can be used to evaluate CV and ΔE_{rel}. It is important to notice that the calculation of the shown uncertainties does not include any systematic error. The bias in measurement, e.g. the temperature measurement on the adsorber casing at only one point, can lead to even higher uncertainties than calculated.

The model provides good simulation accuracy for the calibration measurement. Figure 7.10a shows that the coefficients of deviation (CV) of the heat flow rates are quite low: $CV_{\mathrm{des}} = 8.5\,\%$ for desorption and $CV_{\mathrm{ads}} = 6.9\,\%$ for adsorption. These values attest an excellent accuracy compared to $CV_{\mathrm{adsorber,L}} = 25.3\,\%$ for the heat flow rates of the zeolite adsorber after calibration of Lanzerath's (L) adsorption chiller [93]. The very accurate description of the heat losses ($CV_{\mathrm{loss}} = 3.9\,\%$), which were neglected by Lanzerath [93] in their chiller model, is certainly fundamental here.

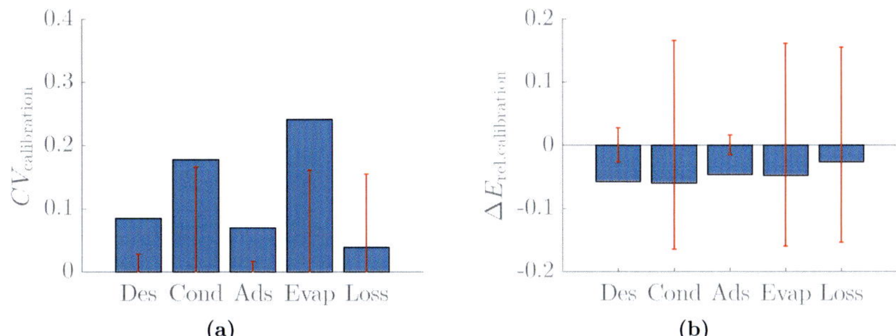

Figure 7.10: (a) Coefficient of variation CV of the heat flow rates (b) and relative deviation of thermal energy ΔE_{rel} of the calibration measurement. Red error bars show the corresponding relative measurement uncertainties.

For the evaporator heat flow rates, the coefficients of variation are $CV_{\mathrm{cond}} = 17.8\,\%$ for condensation and $CV_{\mathrm{evap}} = 24.8\,\%$ for vaporization (cf. Figure 7.10a). These values are obviously higher than for the adsorber, but the measurement uncertainty is also higher in the evaporator than in the adsorber due to lower absolute heat flow rates. The coefficients of performance of the evaporator are comparable to Lanzerath's results [93]: $CV_{\mathrm{cond,L}} = 33.3\,\%$ and $CV_{\mathrm{evap,L}} = 20.2\,\%$. We can conclude that, despite the larger geometry of the TES adsorber compared to the chiller adsorber, the lumped model describes the dynamics of the storage unit very accurately.

The accuracy of the energies transferred during the charging and discharging period, i.e. the relative deviation of thermal energy ΔE_{rel}, is much better than the measurement accuracy of the evaporator and of the heat losses (cf. Figure 7.10b): $\Delta E_{\mathrm{rel,cond}} = 5.9\,\%$, $\Delta E_{\mathrm{rel,evap}} = 4.7\,\%$ and $\Delta E_{\mathrm{rel,Loss}} = 2.6\,\%$. For the energy that is stored and released by the adsorber, the accuracy is in the range of twice the measurement uncertainty: $\Delta E_{\mathrm{rel,des}} = 5.7\,\%$ and $\Delta E_{\mathrm{rel,ads}} = 4.6\,\%$. These values are much better than the simulation accuracy of e.g. Belmonte et al. [105], who achieve relative deviations of 8–13 % between measured and simulated energies during charging and discharging of a latent TES system with paraffin. Consequently, storage performance figures, such as the energy recovery ratio and the energy storage density, can be determined very accurately with the calibrated model of our adsorption-TES unit.

7.3 Model validation for varied storage applications

In the next step, the prediction accuracy of our calibrated model is quantified. For this purpose, we use measurement data which was not used for calibration. The aim is to verify that the storage-unit model, after calibration with a simple charging/discharging cycle, is capable to accurately describe more complex thermal energy storage (TES) processes from various applications.

The validation measurements include the brewing process and residential heating with variations of temperatures and of the discharging procedure: discharging after a storage period (cf. Sections 6.2.3 and 6.2.4) and direct discharge with a two-part discharging period: first, the sensible heat is recovered with the valve closed between adsorber and evaporator, then the valve is opened and adsorption proceeds.

For validation, the calibrated model is used to simulate the storage performance during a storage cycle measurement. The prediction accuracy is then quantified by comparing the simulation to the measurement data. Table 7.4 gives an overview of the measurements that are used for validation of the dynamic model. All experiments were performed with the setup described in Chapter 5. The measurement procedures are explained in the following Sections 7.3.1–7.3.3.

7.3.1 Residential-heating application

In Chapter 6, we describe measurements with discharging temperatures for residential heating and variable storage times τ from direct discharge and 2 h storage time to seasonal storage, and with varying charging temperatures from 175 to 250 °C. Besides the calibration measurement with a discharging temperature of 250 °C and 40 °C in the evaporator, we use direct-discharge measurements with 20 and 40 °C in the evaporator and discharging temperatures of 175–250 °C for model validation.

The measurements with 2 h storage from Section 6.2.3 are also simulated with the calibrated model. In these measurements, the charging and discharging process is delayed by a 2 h storage period. The heat losses during the storage period lead to cooling of the adsorber; the temperature in the adsorber decreases.

Table 7.4: Overview of measurement procedures for storage applications of residential heating and industrial process: charging period with temperature of condensation T_{cond} and desorption end temperature T_{des}; storage times τ and discharging period with temperature of vaporization T_{evap} and adsorption end temperature T_{ads}; heat flow rate limit $\dot{Q}_{\text{ads}}$ to open the valve between adsorber and evaporator during discharge-control measurements. The calibration measurement is underlined in the first row, the rest is used for validation.

	Charging	Storage	Discharging	
Direct Discharge	<u>250 °C</u> $= T_{\text{des}}$, 225 °C, 200 °C, 175 °C; $T_{\text{cond}} = 40\,°\text{C}$		$T_{\text{ads}} = 70\,°\text{C}$; $T_{\text{evap}} = 40\,°\text{C}$	Residential Heating
	250 °C $= T_{\text{des}}$, 225 °C, 200 °C, 175 °C; $T_{\text{cond}} = 20\,°\text{C}$		$T_{\text{ads}} = 70\,°\text{C}$; $T_{\text{evap}} = 40\,°\text{C}$	
2 h Storage	250 °C $= T_{\text{des}}$, 225 °C, 200 °C, 175 °C; $T_{\text{cond}} = 40\,°\text{C}$	$\tau = 2\,\text{h}$	$T_{\text{ads}} = 70\,°\text{C}$; $T_{\text{evap}} = 40\,°\text{C}$	
Discharge Control	250 °C $= T_{\text{des}}$; $T_{\text{cond}} = 40\,°\text{C}$		$\dot{Q}_{\text{ads}} = 2\,\text{kW}$; $T_{\text{ads}} = 70\,°\text{C}$; $T_{\text{evap}} = 40\,°\text{C}$	
	250 °C $= T_{\text{des}}$; $T_{\text{cond}} = 20\,°\text{C}$		$T_{\text{ads}} = 70\,°\text{C}$; $T_{\text{evap}} = 20\,°\text{C}$	
Seasonal Storage	250 °C $= T_{\text{des}}$; $T_{\text{cond}} = 40\,°\text{C}$	$\tau = \infty$	$T_{\text{ads}} = 70\,°\text{C}$; $T_{\text{evap}} = 40\,°\text{C}$	
Direct Discharge	250 °C $= T_{\text{des}}$; $T_{\text{cond}} = 60\,°\text{C}$		$T_{\text{ads}} = 120\,°\text{C}$; $T_{\text{evap}} = 60\,°\text{C}$	Industrial

The measurement for a seasonal storage application is introduced in Section 6.2.4. Seasonal storage represents the extreme case of a storage period with cooldown to environment before discharging.

In all residential heating measurements, charging of the storage unit is immediately started after a discharging period of the same measurement type. All discharging periods are terminated when the adsorber outlet temperature falls below 70 °C. As a result, the measurements do not start from steady-state conditions, but after a sudden change in the fluid flows of the circulation systems after switching from one operation mode to another: from heating to cooling mode and vice versa. The sudden change leads to an overshoot in the controls for temperatures and heat flow rates, which oscillate for a short time.

In the residential heating measurements, the temperature in the water circulation system providing heating and cooling to the evaporator unit is set to 20 °C and 40 °C (cf. Table 7.4), emulating the temperature of the environment or of a solar-thermal unit in winter [147].

7.3.2 Application with discharge control using the valve between adsorber and evaporator

We further validate the model with measurements where the valve between adsorber and evaporator acts as actuator for discharge control: the valve stays closed at the beginning of discharging and is opened when the heat flow rate falls below 2 kW.

Temperatures of the discharge-control measurements are again chosen for a residential heating application: desorption until 250 °C, adsorption until 70 °C and vaporization and condensation at 20 and 40 °C (cf. Table 7.4).

With our experimental setup (cf. Chapter 5) we hereby tested a new operation strategy with the future goal to control the heat flow rate during discharging: the valve between adsorber and evaporator stays closed at the beginning of discharging, until the heat flow rate is decreased to a certain level. When the valve is opened, adsorption starts and the heat flow rate rises again. With the help of a controllable valve, this strategy could be used to stabilize the discharging-heat flow rate, which is still an unsolved issue for adsorption TES application, as discussed in Section 2.1.

7.3.3 Industrial application

The industrial application investigated in Chapter 4 is also used to validate our improved model of the adsorption-TES unit. The validation measurement includes a charging and subsequent discharging period at temperatures of the industrial application (cf. Table 7.4): during charging, the adsorber is heated from 120 °C to 250 °C while heat is released during condensation at 60 °C. Discharging is terminated when the adsorber outlet temperature reaches 120 °C again, while the evaporator is constantly heated to 60 °C.

7.4 Prediction Accuracy

In this section, the prediction accuracy of the model is quantified for the validation measurements from the previous section. The simulation results and the model validity are discussed in Section 7.4.1. To conclude, we compare the prediction accuracy of our model to other models from literature in Section 7.4.2.

We quantify the prediction accuracy by comparing the simulated heat flow rates and transferred energies to their corresponding measured result. The comparison leads to the coefficient of variation CV and the relative deviation of thermal energy ΔE_{rel} (cf. Section 7.2.3). The prediction accuracy is high for low values of the coefficient of variation CV and of the relative deviation of the transferred energy ΔE_{rel}.

Figure 7.11 shows the coefficient of variation CV, which is our measure for the prediction accuracy of the heat flow rates. The CV is plotted together with the measurement uncertainty of the heat flow rates (cf. Equation (2.5)) for all measurements listed in Table 7.4. The prediction accuracy of the transferred energies is reflected by the relative deviation of thermal energy ΔE_{rel}, which is shown in Figure 7.12, also in comparison to the measurement uncertainty.

The coefficient of variation CV and the relative deviation of thermal energy ΔE_{rel} for desorption and adsorption describe the prediction accuracy of the adsorber performance. During charging, the desorption heat flow rates (CV_{des}) as well as the transferred energies ($\Delta E_{\text{rel,des}}$) are very accurately described by the model in all measurements, with deviations of less than 15 %, cf. Figure 7.11 and Figure 7.12.

The coefficients of variation of the adsorption heat flow rates CV_{ads} are higher the more the adsorber is cooled down after desorption, before the valve to the evaporator is opened.

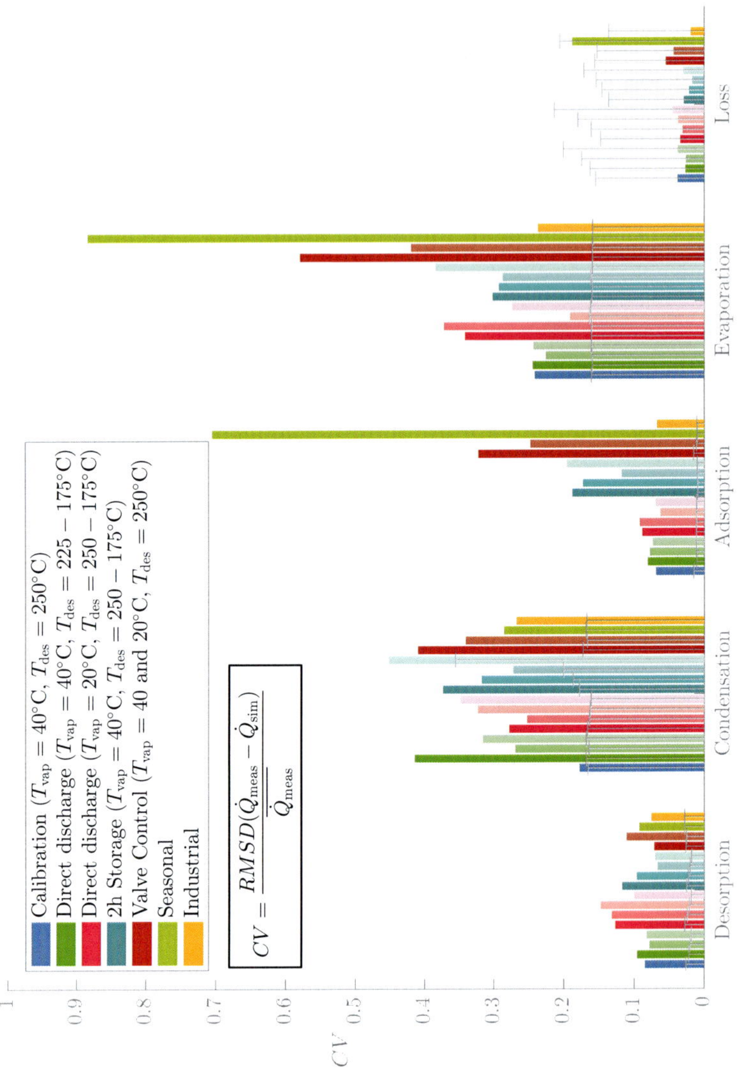

Figure 7.11: Coefficient of variation CV of the heat flow rates (cf. Equation (2.5)) during desorption, condensation, adsorption, vaporization and of the heat-loss rates for all measurements listed in Table 7.4, plotted with the average relative measurement uncertainty $U_{rel}^{95\,\%}$ of the corresponding heat flow rate.

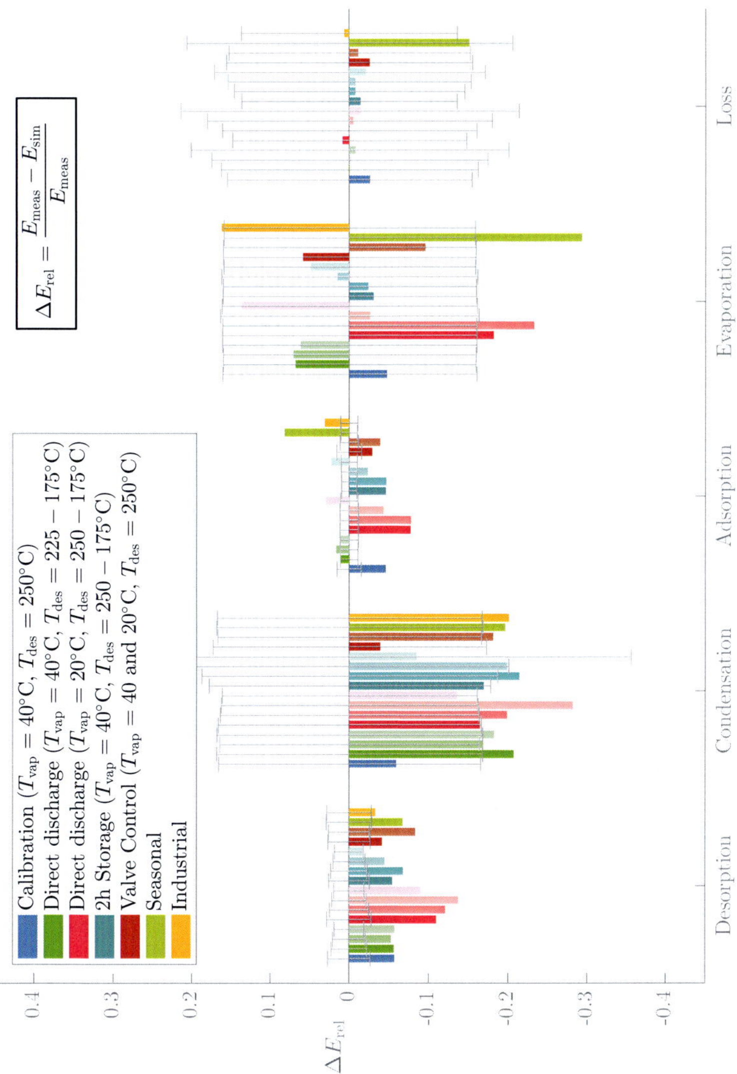

Figure 7.12: Relative deviation of thermal energy ΔE_{rel} during desorption, condensation, adsorption, vaporization and relative deviation of the total heat loss for all measurements listed in Table 7.4, plotted with the relative measurement uncertainty $U_{rel}^{95\,\%}$ of the corresponding energy.

While the CV_{ads} of the direct-discharge measurements and the industrial measurement is in the same range as for calibration, i.e. below 10 %, it increases to 20 % for the measurement with a 2 h storage period, 32 % for the discharge-control measurement and even 70 % for seasonal discharging. During seasonal discharging, the dynamics of the adsorber are difficult to describe due to large gradients. Nevertheless, the transferred thermal energy during adsorption is predicted very accurately (cf. Figure 7.12), even during seasonal discharging with $\Delta E_{\text{rel,ads}}$ between 1 and 8 %.

The prediction of the evaporator performance is evaluated by the coefficient of variation CV and the relative deviation of thermal energy ΔE_{rel} for condensation and vaporization. These measures are generally larger than for the adsorber due to lower evaporator heat flow rates and thus larger relative measurement uncertainties, cf. Figure 7.11 and Figure 7.12. As the deviations of the transferred energy ΔE_{rel} during condensation and vaporization are close to the measurement uncertainty, we can state that the evaporator performance is accurately described by the model. Deviations of the transferred energies in the evaporator, especially during condensation, are larger than in the adsorber, but still within a range of $\Delta E_{\text{rel,evap/cond}} < 30\,\%$.

Throughout all measurements, the heat losses are very accurately described by the model and deviations are within the measurement uncertainty, both for heat-loss rates (cf Figure 7.11) as well as for the total heat loss (cf. Figure 7.12).

7.4.1 Discussion of model validity

In this section, we discuss the prediction accuracy, the remaining deviations between simulation and measurement, and the model validity of our adsorption-TES unit model. To support the discussion, simulation results for selected measurements are shown. For all other validation measurements, simulation results are given in the Appendix C.2.

As in the calibration measurement (cf. Section 7.2.2), in all validation measurements, the heat flow rate is slightly underestimated at the beginning of each de- and adsorption phase and slightly overestimated at the end. Even the time span of underestimation in the first quarter of an hour of discharging is equal to the calibration measurement (cf. Figure 7.9, bottom). The same reasons apply as for the calibration measurement (cf. Section 7.2.2): the lumped model approach flattens the sharp increase of heat flow rates at the beginning of each charging and discharging period and the adsorption characteristics underestimate the water affinity of the zeolite at low loadings and thus the heat release at the beginning of adsorption. This mismatch of discharging-heat flow rates is more pronounced, the colder the adsorbent is and hence the larger the driving force for adsorption should be.

Adsorption characteristics

To substantiate this explanation, we take a look at the simulation result of the discharge-control measurement: Figure 7.13 shows the resulting temperatures and heat flow rates. While the valve is closed at the beginning of discharging, no adsorption or vaporization

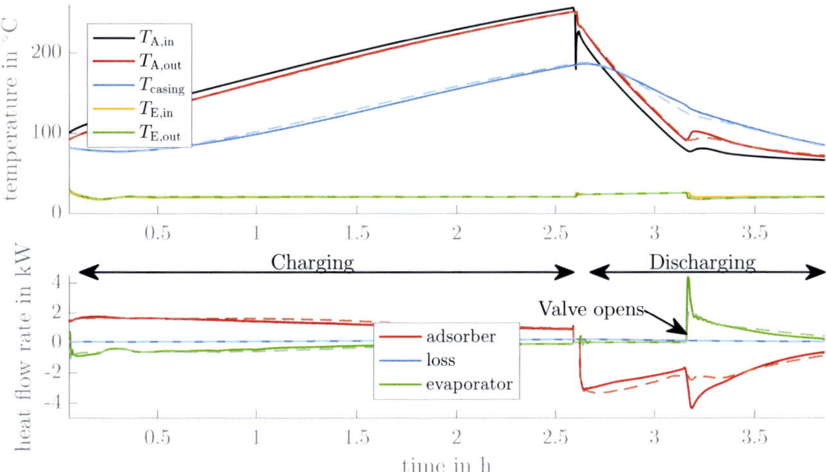

Figure 7.13: Validation measurement for residential heating with closed valve at beginning of adsorption, $T_{des} = 250\,°C$ and $T_{evap} = 20\,°C$. Top: temperatures of adsorber and evaporator. Bottom: heat flow rates of adsorber, evaporator and losses to the environment. Dashed lines show the corresponding simulation results.

occurs, but the adsorber faces a purely sensible change of internal energy. In this case, the simulated discharging-heat flow rate of the adsorber is rather determined by the description of heat transfer and geometry than by the adsorption characteristics. The simulation slightly overestimates the heat flow rate of the adsorber at the beginning of discharging when the valve is closed (cf. Figure 7.13, bottom). As the closed valve inhibits adsorption, we can exclude any influence of the adsorption characteristics. Thus, the overestimated heat flow rate is due to the heat-transfer coefficient, which is slightly overvalued. We can now confirm the assumption that the heat-transfer coefficient in the adsorber is overvalued by calibration: during calibration, an elevated heat-transfer coefficient compensates for the underestimated heat release by the adsorbent at the beginning of adsorption (cf. discussion in Section 7.2.2).

The underestimation of the heat release at the beginning of adsorption, i.e. at low water loadings, is further visible as soon as the valve is opened. Just before the valve is opened, the adsorber unit is already cooled down by 150 K and is very dry. As soon as the valve is opened, the measured heat flow rate increases sharply. This heat flow rate is strongly underestimated by the model due to adsorption characteristics that are presumably underestimated at low water loadings. We have further confirmed that the model of the adsorption characteristics influences the simulation result, because both evaporator and adsorber heat flow rates are underestimated by the model at the beginning of each adsorption.

As a result, the dynamics of the adsorption process are more difficult to describe the colder the adsorber is. Hence, the least accurate prediction of the adsorption dynamics is shown by the simulation of the validation measurement of seasonal discharging (cf. CV_{ads} in Figure 7.11).

As described in Section 6.2.4, for seasonal discharging, we heat the evaporator to 40 °C, open the valve to the adsorber and wait until the temperature of the adsorber surface exceeds 70 °C before turning on the oil pump to recover the discharging heat. Figure 7.14 shows the simulation results for seasonal adsorption.

The simulation follows the trends of the measurement, including the two peaks and the drop in the adsorber heat flow rate at the beginning of discharging. Nevertheless, we again see the underestimation of heat flow rates in the first minutes, even leading to a minor inversion in flow direction of the adsorber heat flow rate (cf. Figure 7.14). This inversion in the simulation is due to the temperature of the adsorber being below the temperature of the input oil flow, an inversion in driving temperature difference. Reasons might again be an underestimation of adsorption at low water loadings in the adsorber, confirmed by the evaporator heat flow rate: when the valve is opened after cooldown to environment, the measured peak heat flow rate in the evaporator is predicted lower by the simulation. Nevertheless, the released energy by the adsorber is in good agreement to the measured results with deviations below 10 % (cf. Figure 7.12).

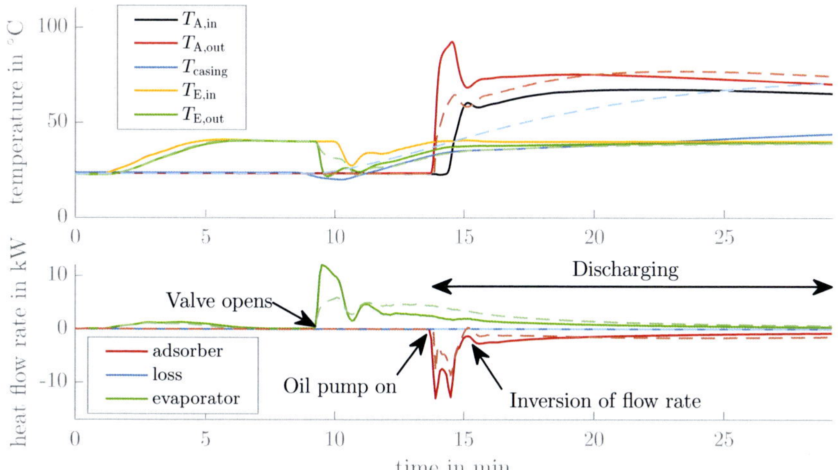

Figure 7.14: Validation measurement of discharging from seasonal application, $T_{des} = 250\,°C$ and $T_{evap} = 40\,°C$. Top: temperatures of adsorber and evaporator. Bottom: heat flow rates of adsorber, evaporator and losses to the environment. Dashed lines show the corresponding simulation results.

Lumped-model approach

The simulation results for seasonal adsorption also reveal a limitation of the lumped-model approach. In the measurement, adsorption first takes place at the outer beads. The metal lamellae in the adsorber conduct the heat to the heat exchanger tube, leading to a sharp increase in the adsorber heat flow rate. As the adsorber is modeled as one homogeneous block that cannot heat up as fast as the outer beads in the measurement, the heat flow rate is vastly underestimated at the beginning of seasonal adsorption, (cf. Figure 7.14, bottom). The same applies for the validation measurement with discharge control where the gradient over the adsorber bed is presumably large when the valve is opened during discharging (cf. Figure 7.13, top). As a result, the coefficients of variation of the adsorber heat flow rates are larger for the discharge-control and the seasonal measurements compared to the other validation measurements (cf. Figure 7.11).

The underestimated heat flow rates at the beginning of both de- and adsorption phases in all validation measurements can also be traced back to the lumped-model approach. In all measurements, thermal energy in the metal parts can be recovered fast at the beginning of adsorption; metal parts take up heat fast at the beginning of desorption. However, the lumped model smooths these heat flow rates. Though, these deviations in heat flow rates do not lead to deviations in transferred energy. The overestimated heat flow rate at the end of a process compensates for the underestimation at the beginning and the transferred energy is the same in the end. Thus, the total amount of thermal energy is accurately predicted, but with less dynamics, i.e. ΔE_{rel} (Figure 7.12) is always better than CV (Figure 7.11).

Adsorptive water influencing the casing temperature

In the measurement for seasonal discharging, we perceive that the vaporization in the evaporator starts with intense boiling due to the large driving force for adsorption with the dry adsorber cooled down to ambient temperature. Figure 7.14 (top) shows the corresponding temperatures in the storage unit. The temperature evolution of the adsorber casing is obviously influenced by cold water vapor from the evaporator. However, the model does not follow this trend, since the water vapor flow near the casing wall is not described by the model. The influence of cold water splashing from the evaporator can also be observed in the measurement in Figure 7.13 (top), when the valve between evaporator and adsorber is opened.

Incondensable gases

Regarding the deviation in energy, the validation shows excellent accuracy of the model, except for condensation. We see a systematic overestimation of the simulated energy of desorption and condensation, cf. Figure 7.12. During condensation, a small amount of incondensable gases within the storage unit containment already hinders the condensation process in the measurement [148]. The model does not cover these effects, leading to an overestimation of the simulated energy during condensation.

Excellent representation of heat losses

The prediction of the heat losses is very accurate for all validation measurements: deviations are mostly smaller than 5 %. Even for the extreme case of cooldown to environment during a seasonal measurement, the accuracy of the heat-loss rate, which refers to the charging, discharging and the complete cooldown phase, is very good and the CV_{loss} is within the measurement accuracy, cf. Figure 7.11.

The temperatures and heat flow rates of the charging period and subsequent cooldown of a seasonal measurement are shown in Figure 7.15. During the cooldown phase, the

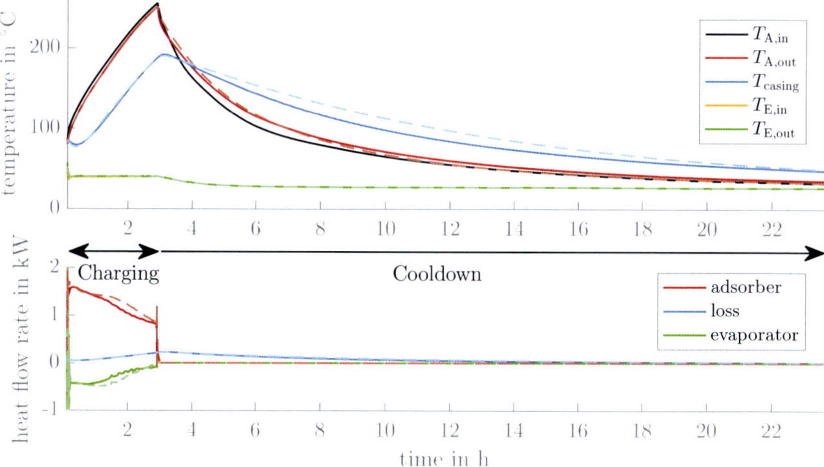

Figure 7.15: Validation measurement of charging and cooldown from seasonal application, $T_{des} = 250\,°C$ and $T_{cond} = 40\,°C$. Top: temperatures of adsorber and evaporator. Bottom: heat flow rates of adsorber, evaporator and losses to the environment. Dashed lines show the corresponding simulation results.

temperature of the adsorber casing decreases more slowly than the temperatures in the oil circulation system, due to poorer insulation of the oil pipes. Both temperatures are accurately predicted by the model.

In general, all validation measurements (cf. Figures 7.13–7.17) prove excellent agreement between simulated and measured heat-loss rates.

Discharging after storage periods

From the discussion of the simulation results of the seasonal measurements, we can conclude that the model excellently describes the heat losses during cooldown to the environment. As a result, the model very accurately predicts the energy output of the adsorption-TES unit after the cooldown.

For short storage periods, the prediction accuracy of our model is even better, in particular when predicting the dynamics of adsorption TES: Figure 7.16 (bottom) shows the heat

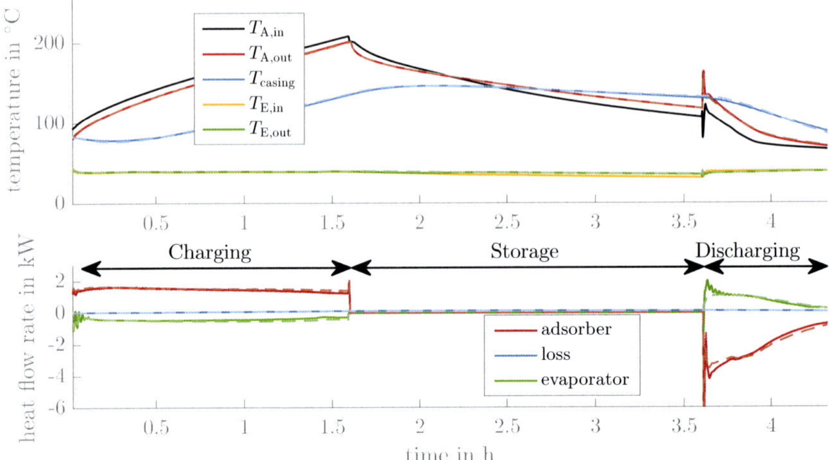

Figure 7.16: Validation measurement with 2 h storage time, $T_{des} = 200\,°C$ and $T_{evap} = 40\,°C$. Top: temperatures of adsorber and evaporator. Bottom: heat flow rates of adsorber, evaporator and losses to the environment. Dashed lines show the corresponding simulation results.

flow rates during one complete cycle including a 2 h storage period. The simulation results (in dotted lines) match the measurement data very well, even after the storage period. In particular, the simulated temperatures and the heat losses excellently match the measurements. We can conclude that the model very accurately predicts the performance of the adsorption-TES unit for an application with short storage periods (cf. Figures 7.11, 7.12 and 7.16.)

Storage unit performance at different temperatures

Our last test for the model is to predict the performance of a storage cycle with temperatures that differ from the conditions of the calibration measurement. The validation measurement for the industrial application emulates the brewing process conditions from Chapter 4: the evaporator is at 60 °C and discharging heat is released at 120 °C. Figure 7.17 (top) shows that the temperatures are very accurately predicted by the storage-unit model, which was calibrated with a measurement at lower temperatures, cf. Figure 7.9, top. As we saw

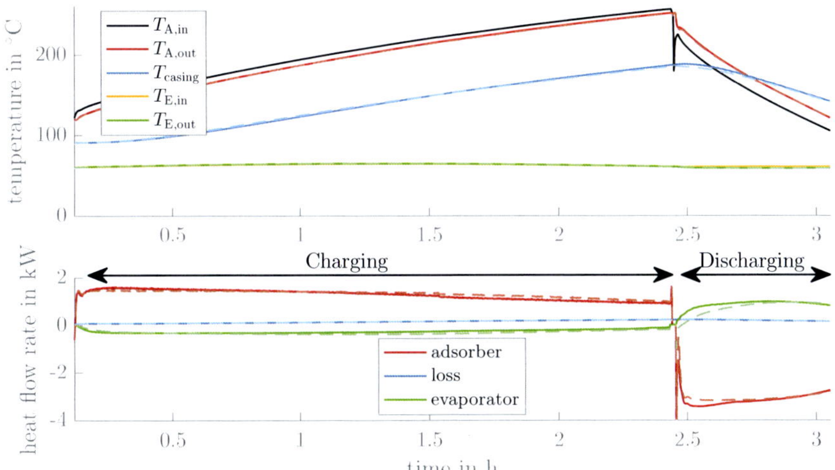

Figure 7.17: Validation measurement of industrial application, $T_{des} = 250\,°C$, $T_{ads} = 120\,°C$ and $T_{evap} = 60\,°C$. Top: temperatures of adsorber and evaporator. Bottom: heat flow rates of adsorber, evaporator and losses to the environment. Dashed lines show the corresponding simulation results.

already in Section 6.1, the pressure in the storage unit does not influence the heat losses of the storage unit to a large extent.

The heat flow rates are also very accurately predicted (cf. Figure 7.17, bottom) despite the trend of underestimated heat flow rates at the beginning of each charging and discharging period and overestimation at the end. However, the high prediction accuracy for the industrial application measurement is evident in the coefficient of variation (cf. Figure 7.11) and the relative deviation of thermal energy (cf. Figure 7.12), which are particularly low for the charging heat (desorption) and the process heat (adsorption). From the validation results of the industrial application measurement (cf. Figure 7.17), we conclude that the model very accurately predicts the measurement with temperatures that differ from calibration.

To conclude this section: the model is valid for a broad range of process conditions and describes the performance of our adsorption-TES unit very accurately, in particular for short storage periods. The results of the thorough model validation show the general applicability of our adsorption-TES model.

7.4.2 Comparison to the model of Lanzerath [93]

The quality of our adsorption-TES model can be evaluated by comparing the prediction accuracy to other models from the literature. As emphasized in Section 2.5.5, there are only few models with given accuracy measures to compare to. Lanzerath [93] gives the prediction accuracy of his zeolite-chiller model with an overall coefficient of variation $CV_{\text{overall,L}} = 27.3$–$44.7\,\%$. This $CV_{\text{overall,L}}$ is the arithmetical mean of the CV of the adsorber and evaporator heat flow rates.

In comparison to Lanzerath [93], our model provides an excellent prediction accuracy of the adsorber heat flow rates and the heat losses: the CV_{des} of the heat flow rate during desorption is always below 15 %. The direct-discharge simulations yield a CV_{ads} lower than 10 %, after 2 h, the CV_{ads} is the still lower than 20 % and the measurements with a closed valve at the beginning of discharging are predicted with less than 32 % deviation (cf. Figure 7.11). The comparison to the model from Lanzerath [93] also confirms the very good prediction accuracy of our adsorption-TES model for the evaporator heat flow rates during condensation and vaporization: we achieve a prediction accuracy for the evaporator that is mostly in the range of the results from Lanzerath [93] with $CV_{\text{cond}} = 25 - 45\,\%$ and $CV_{\text{evap}} = 18 - 41\,\%$, if we exclude the discharge-control and seasonal measurement (cf. Figure 7.11).

The prediction accuracy of the total energies can be compared to the prediction accuracy of the coefficient of performance, which is 18.7 % in average for the zeolite-chiller of Lanzerath [82]. Our model achieves an excellent prediction accuracy with small relative deviations of the transferred thermal energies ΔE_{rel}, ranging from 2–14 % during desorption, 1–8 % during adsorption and mostly below 3 % for heat losses, as given in Figure 7.12. ΔE_{rel} reaches slightly higher values of 4–28 % during condensation, which is 17.5 % in average, and 1–29 % during vaporization, with an average value of 10 %.

The comparison to the prediction accuracy of the chiller model reveals that the calibration of the adsorption TES yields a very accurate model despite the much larger adsorber geometry. We can thus conclude that the accuracy of a lumped-model with 1-D discretized heat exchangers is sufficiently high for our adsorption-TES system. The validated adsorption-TES model allows extensive simulation studies.

7.5 Summary

We present an experimentally validated, dynamic model of an adsorption thermal energy storage (TES) unit with a valid description of heat losses. Heat losses are determined from steady-state measurements (Chapter 6). Measurements of the surface temperature of the adsorber casing allow for separate calibration of the heat transfer from the adsorber to the environment.

To describe the storage unit, we choose a lumped-parameter model with 1-D discretization of heat exchangers and few heat and mass transfer resistances. Despite these simplifications of the adsorber geometry, the calibration yields a very accurate model. The reduced model complexity of the lumped-model approach enables system simulations.

We validate the model for process conditions different from calibration: the model is used to predict the storage performance of the validation measurements. For validation, we choose measurements from both the industrial process analyzed in Chapter 4 and residential heating processes with various storage times (cf. Chapter 6) and with a discharge control that uses the valve between evaporator and adsorber.

With the various validation measurements, we prove that we provide a versatile model which is capable of describing the storage unit performance for different process conditions. The accurate description of heat losses is crucial to predict storage behavior. The validation results show that the adsorption-TES model properly predicts the storage behavior for storage periods with cooling of the storage unit, because heat losses of the thermal storage unit are accurately represented by the calibrated model.

We show that the model provides good prediction accuracy despite strong simplifications of the geometry. The lumped model assumes homogeneous temperatures and loading, which does not perfectly describe the complex geometry of the adsorber. Strongly transient heat flow rates are smoothed, in particular at the beginning of each charging and discharging period. However, with few exceptions, the predicted heat flow rates precisely follow the trend of the measured results.

During calibration of the heat-transfer coefficient in the adsorber, a compromise is found between the high heat flow rates at the beginning of each phase and the slower heat release later. Thus, the heat flow rate at the beginning of adsorption is slightly underestimated by the model, both in the calibration and all validation measurements. In particular, the model underestimates the discharging-heat flow rates for measurements where the adsorber is very cold at the beginning of adsorption.

In general, deviations in heat flow rates do not necessarily lead to deviations in transferred energy. An underestimated heat flow rate at the beginning of a process can compensate for overestimation at the end. In such case, the total amount of thermal energy is accurately predicted, resulting in a high prediction accuracy for storage performance measures like storage efficiency or energy density.

The deviations between measurement and simulation are quantified with two accuracy measures, one for the heat flow rates and one for the energy in- and output of the storage unit. The proposed accuracy measures can be used to compare different mathematical models regarding their suitability to describe the storage performance.

Unfortunately, comparison to literature is difficult, since the deviations between simulation and measurement of adsorption-TES systems are not quantified by other authors, except for visual comparison. Compared to a well-described model of an adsorption chiller, we can confirm that our storage-unit model achieves high prediction accuracy.

To sum up, the evaluation of the simulation results for the residential heating and the industrial storage application confirms the assumption that the storage model can be used to describe various storage applications. Despite some remaining deviations between measurements and simulations, the lumped model reaches a high prediction accuracy. We thereby prove that accurate simulations of storage applications on a system level are possible with our dynamic, lumped-parameter model.

8 Conclusions and Outlook

This thesis proposes methods to evaluate adsorption thermal energy storage (TES) in the context of residential and industrial applications. First, an experimental and model-based approach is used to assess the benefits of adsorption TES for a new application. Second, the performance of our adsorption-TES unit is experimentally analyzed. The thesis finally provides a model of this storage unit that validly describes and predicts the storage performance.

In this chapter, we conclude the achievements of this thesis (Sections 8.1–8.3) and propose further steps towards the application of adsorption TES in practice (Section 8.4).

8.1 Assessment of a new application for adsorption thermal energy storage

This thesis provides a thorough assessment of adsorption TES integrated for the first time in an industrial process. The literature review in Section 2.3 reveals that the evaluation of adsorption TES for new applications requires (1) the analysis of the storage performance in the system context and (2) the comparison to alternative technologies. To allow for a thorough analysis in Chapter 4, we develop a sound dynamic model for our small-scale adsorption-TES unit. The model is used to study and compare the performance of adsorption TES combined with cogeneration for the energy supply of a batch process. Beer brewing is considered as an example of an industrial batch process. As benchmark, we consider both a peak boiler and latent TES.

The study shows that adsorption TES has the potential to increase energy efficiency significantly. The primary energy consumption of the process can be reduced by up to 25 % compared to benchmark technologies. However, successful integration of adsorption TES requires the appropriate integration of low-grade heat: preferably, the low-grade heat is demanded while the storage unit is charged and available at times of discharging. In conclusion, an energy supply system with adsorption TES is beneficial when it makes use of the heat pump effect. Thus, adsorption TES yields advantages if it is applied to a batch process with waste heat and with heat demands on several temperature levels.

8.2 Experimental analysis of storage performance

The results of Chapter 4 emphasize the impact of heat losses on the performance of TES systems (cf. Sections 2.4 and 4.6). These heat losses need to be quantified to apply storage in practice. In Chapter 6, we demonstrate that heat losses can be accurately determined from steady-state measurements. The heat-transfer coefficients to the environment of the storage unit are low due to the low pressure inside the adsorber casing and due to the insulation of the casing. The low heat transfer leads to high storage efficiencies: the charging efficiency is above 91 % and discharging efficiency is higher than 95 %.

We further quantify the energy recovery ratio and the energy storage density for varying charging temperatures and storage times ranging from simple storage cycles with direct discharge to seasonal-storage cycles with discharging after cooldown to the ambient temperature.

For seasonal storage, the heat losses strongly reduce the storage density. In particular during discharging, reheating of the storage unit to discharging temperature requires a large part of the stored thermal energy. Thus, for a seasonal storage application, a low temperature of discharging would be beneficial.

For short-term storage, the energy storage density can be enhanced significantly by charging at high temperatures. While the energy recovery ratio only decreases slightly with charging temperature due to heat losses, the energy storage density increases more strongly with temperature. The measured energy storage density during continuous operation corresponds to 93 % of the theoretical maximum for lossless operation and to 76 % for 2 h storage time. The results show that adsorption TES can provide for efficient storage.

We recommend to choose a high maximum storage temperature for short-term storage applications. For process conditions similar to the experimental conditions in this thesis, the high gain in energy storage density compensates for the slight reduction of the energy recovery ratio. Consequently, adsorption TES systems can be beneficial, e.g. in industrial processes.

8.3 Experimental model validation

To exploit the benefits of adsorption TES systems for efficient energy systems, reliable models are needed. The survey of recent literature in Section 2.5 reveals that the simulation of adsorption TES systems requires valid models for the heat losses and for the system dynamics.

In Chapter 7, we provide a valid dynamic model for our adsorption-TES unit with a thorough description of heat losses. The complex geometry of the adsorber is simplified with a lumped-parameter description. The dynamic model is calibrated with experimental data. We propose a simple calibration procedure that requires only few measured values and yields a very accurate description of the dynamics of our adsorption-TES unit. The description of the storage performance is excellent in comparison to models from literature.

Moreover, a comprehensive validation of the adsorption-TES model is provided, based on validation measurements with various temperatures, various storage times and a discharging control using the valve between adsorber and evaporator. The calibrated model accurately predicts the temperatures and heat flow rates of our adsorption-TES unit for these various process conditions.

However, our validation results show that strong transients cannot be perfectly represented by a lumped model of a storage unit with a considerable thermal mass and low thermal conductance. Thus, for larger storage devices and for storage applications with strongly fluctuating heat flow rates, we assume that a spatially extended model could become necessary to accurately describe the dynamics of the storage system, in particular for seasonal storage. Nevertheless, with our storage unit and the short-term applications it is designed for, we achieve a very good accuracy to predict the system dynamics with our lumped model.

We furthermore quantify the prediction accuracy to make the model quality measurable and comparable to other models. The simulated adsorber heat flow rates are predicted with deviations of 7 %–15 % during charging and 6 %–9 % at direct discharging. After short storage periods, the deviations are still below 20 % for discharging. The evaporator is predicted with deviations of 18 %–45 % and the heat losses deviate by less than 5 % from our measurements. The deviations in the predicted energy in- and outputs are below 14 % in the adsorber and below 30 % in the evaporator. The total heat loss is predicted with negligible deviations of less than 3 %, except for the seasonal measurement with 15 %. The prediction accuracy of our model thus reaches better values than an adsorption chiller model from literature (cf. Section 7.4.2). Unfortunately, we could not find any adsorption-TES models with given accuracy to compare to.

We conclude that the dynamic, lumped-parameter model provides sound predictions and describes heat losses with excellent accuracy. The model thus enables a simulation-based analysis of storage applications, such as industrial applications and heating applications with storage periods of some hours.

8.4 Future Research

The outcome of this thesis can be further applied to exploiting the advantages of adsorption thermal energy storage (TES) for efficient future energy supply systems. Recommendations on future research are provided in the following.

8.4.1 System analysis for further applications

To exploit the advantages of adsorption TES systems, the applications should benefit from the unique features of the storage technology. Furthermore, the performance of adsorption TES should outperform the alternative technologies for each application. In this section, some ideas for applications of adsorption TES are outlined.

In particular in residential heating applications, it is difficult to outperform a sensible TES system with water in terms of cost and system simplicity. Compared to water, adsorption TES has the advantage of long-term storage and of the heat pump effect. However, mere long-term or seasonal energy storage cannot be advised for economic reasons [36]. In case of a short-term storage application in residential heating, making use of the heat pump effect would be beneficial by integrating solar thermal energy at very low temperatures to produce discontinuous heating by adsorption TES.

Water might only be outperformed if the aspect of long-term storage can be combined with short-term storage, as Schmidt and Linder [149] propose: they suggest to regenerate a thermochemical storage system at elevated temperatures with surplus electricity and to use the condensation enthalpy to provide hot water to households all year long. The large amount of thermal energy that can be stored for a long time in thermochemical reactions could thus be used in winter for heating purpose.

As we have seen in Chapter 4, thermal energy storage may offer advantages for industrial processes. In particular, we recommend to search for industrial batch processes to apply adsorption TES. Even for large plants, where direct heat recovery would be possible, direct heat recovery is often not implemented to avoid making the subsequent processes dependent on each other [71]. On the contrary, TES systems enable to establish heat recovery without disturbing the chronological sequence of the batch processes.

The unique advantage of the adsorption TES technology is that it makes use of waste heat for process energy supply. On the other hand, the low temperature heat during condensation should also be used, just as N'Tsoukpoe et al. [150] conclude. If the heat pump effect was used, adsorption TES could be very beneficial to increase energy efficiency of industrial batch processes. Once an appropriate application is found, the model proposed in Chapter 7 can be used to optimize the storage integration into the process.

Concerning alternative storage technologies, de Boer et al. [71] criticize that latent TES materials are not yet commercially available for temperatures between 120-200 °C and that heat transfer rates are the most limiting aspect to apply latent TES for industrial processes. Other storage media for this temperature region, such as concrete, need very large storage volumes to achieve the necessary storage capacity. For very short cycle times and heat high flow rates, water vapor storage systems (Ruths storage) are commercially available. To compete with these technologies, the high energy storage density of adsorption systems should be combined with high heat transfer rates to apply adsorption TES to new applications with short cycle times.

Last but not least, to evaluate adsorption TES for new applications, the costs should be considered as well. The costs of a TES system depend on the costs per installed capacity, the energy costs and the number of cycles per year (cf. Section 2.2). All three factors will improve with short cycle times: processes with short cycle times often involve less installed capacity and a higher number of cycles per year. Moreover, short cycle times reduce heat losses (cf. Section 6.2.2), which increases the energy recovery ratio and hence reduces the energy costs and the installed capacity. Thus, our suggestion to search for a new application with short cycle times satisfies the concern of cost efficiency.

8.4.2 Intermittent operation and advanced control of adsorption thermal energy storage systems

The operation strategy of a storage system has to meet the requirements of the chosen application (cf. Section 2.1). One example is thermal energy storage to provide flexibility for electrical load shifting with cogeneration, as Fehrenbach et al. [49] propose. This applications requires flexible operation of the storage unit. To provide this flexibility, the storage unit is charged and discharged gradually with interruptions, or it is even only partially discharged. Thus, the storage unit performs cycles with intermittent operation.

To engineer an adsorption TES system for an application that demands flexibility, the behavior of adsorption TES during intermittent operation should be investigated. In contrast, most experimental investigations so far perform full cycles, e.g. References [32, 33, 74, 101]. In contrast to full cycles, intermittent operation means that the storage unit is discharged partially or discontinuously. Michel et al. [31] include one cycle with intermittent operation in their investigation of an open thermochemical system for seasonal TES: they stop the discharging process in-between and let the system cool down to ambient temperature again before completely discharging the storage unit. However, their reaction system has very slow kinetics and operates below 35 °C, thus the interruption during discharging does not lead to any detectable change in performance.

For an adsorption TES system, an interruption of the discharging process could lead to undesired effects: once discharging is started, the adsorption already takes place and the storage temperature rises. If the heat is not used immediately, the rising temperature causes increasing heat losses. These self-discharging effects have to be addressed by future research on adsorption TES systems.

Another issue with operation strategies of adsorption TES systems is the control of the output power, as discussed in Section 2.1. Yu et al. [30] propose to control the power output by regulating the flow rate of the heat transfer fluid with a variable frequency pump. This approach will certainly work for full storage cycles. However, in combination with intermittent operation, i.e. partial cycles, a more sophisticated control strategy could be beneficial to avoid the self-discharging effect of the adsorption TES systems. For this purpose, we need to regulate the water vapor pressure inside the storage unit (cf. Section 2.1).

We propose a control strategy which offers advantages for cycles with intermittent operation: the valve between adsorber and evaporator could be used to regulate the adsorption progress and thus the heat output of the adsorber. In contrast to controlling the output power with the heat transfer fluid flow rate, the valve influences the discharging process from inside the storage unit. When the valve is closed at the beginning of discharging, the sensible part of the stored energy can be recovered first while the latent part of the energy remains stored.

We have started to test the control strategy with measurements of a simple discharge-control cycle, as addressed in Section 7.3.2 and shown in Figure 7.13: the valve stays closed at the beginning of discharging and the sensible part of the stored energy is recovered first. The valve is then opened as the discharging heat flow rate decreases below the desired heat flow rate. Due to an uncontrollable valve, the heat flow rate increases sharply as soon as the valve is opened. However, a valve that opens gradually could be used to control the water vapor pressure provided to the adsorber. As a result, the adsorption progress and thus the heat output could be controlled. Such a control system with a controllable valve could be in focus of further experimental investigations of adsorption-TES units.

8.4.3 Optimization of component design

Lumped-parameter and 1-D discretized models of adsorption TES are not only applicable to system analysis. These models, when accurately calibrated to measurement data, can give hints if a component might be hindering the charging or discharging process. For instance, the model calibration in Section 7.2.1 reveals that the mass transfer is not limiting the storage process. With a sensitivity study of the heat-transfer coefficients, the limits of further enhancements in heat exchanger optimization could be identified.

Moreover, simple design choices can be analyzed with lumped-parameter 1-D heat exchanger models [87, 151]. One example would be to optimize the energy recovery ratio under space restrictions by changing the size of adsorber and evaporator. However, a more thorough optimization of the storage unit geometry requires 3-D models and a physical description of heat and mass transfer [83]. A geometry-based modeling of adsorbers is particularly necessary for storage systems with strongly transient heat flow rates (cf. Section 7.4.1). To achieve high charging and discharging rates without significant loss of storage capacity, improvements in heat exchanger design are required [71].

8.4.4 Enhanced thermochemical materials

To further develop thermal energy storage systems, it is common consensus that we need enhanced materials with high storage densities. In particular, thermochemical TES systems based on solid-gas reactions are in focus of current research [152]. Numerous studies based on uptake measurements have been conducted to compare novel materials, e.g. References [150, 153, 154], and to find the most promising candidates for thermochemical energy storage.

Drawbacks of these materials are the low thermal stability of the materials and the slow kinetics of the reaction [47]. Richter et al. [155] demonstrate heat transformation with a closed system of calcium chloride and water. They discuss problems with the material stability and agglomeration of the salt. Fopah Lele et al. [83] solve the agglomeration issue of pure salts by using an aluminium honeycomb heat exchanger with strontium bromide. However, the expected performance could not be achieved with the pure salts. To enhance the performance, they refer to composite materials with hydrophilic salts, as proposed by Mauran et al. [156], Yu et al. [157], Zhao et al. [158], Kariya et al. [11] and Myagmarjav et al. [159]. Nonnen et al. [160] experimentally quantify the energy storage density of a seasonal storage unit with a composite adsorbent in open system mode. They achieve higher energy storage densities than with pure zeolite, but only at high partial water pressures, which in winter is an issue for open systems due to very low humidity. Thus, the performance of these enhanced materials again depends on the process conditions.

When chemists have overcome the stability problem for composite materials, applying these to thermal energy storage units offers potential for efficient and compact storage. In particular, closed adsorption storage systems should be investigated, because they offer limited heat loss and do not depend as much on environmental conditions, such as humidity.

However, without an appropriate application, the choice of material and storage geometry will be useless. We thus propose to implement the equilibrium data of novel materials to our model-based analysis to gain first estimates of their performance on a system level. With the validated model from Chapter 7, we have provided the basis to proceed in this direction.

A Experimental Equipment and Material Data

This appendix provides information on the used adsorbent material and on the thermal oil that is used as heat-transfer fluid in the adsorber heat exchanger. Manufacturer data are given, which are used to describe the material properties in the dynamic models.

A.1 Adsorbent

The adsorbent material is a zeolite type 13 X MOLSIV Adsorbent beads by UOP LLC from Des Plaines, IL, USA. The chemical formula of the zeolite is $Na_x\,[(AlO_2)_x(SiO_2)_y]z\,H_2O$. The spherical beads have a diameter of 1.4 mm and a bulk density of 641 kg/m^3.

The crystalline structure of zeolites can be damaged under extreme hydrothermal stress [161]. Unfortunately, the adsorber of our storage unit faced a severe overheating under very high pressure of water vapor due to mishandling during commissioning of the experimental setup. This hydrothermal stress resulted in a reduction of the adsorption capacity of the zeolite. To determine the reduced water uptake of the zeolite, equilibrium measurements were performed as follows:
To prepare for the measurement, the zeolite was completely dried by heating the adsorber to 250 °C while applying a vacuum pump for 24 h. The evaporator was then filled with 2.8 kg of water which was partly adsorbed by the zeolite. The thermal energy storage (TES) system was further heated to steady-state conditions for some hours, to ensure equilibrium. Then, the valve to the evaporator was closed to maintain the water loading in the adsorber. The remaining water content in the evaporator was then measured with a scale. The difference between the 2.8 kg of water and the remaining water in the evaporator equals the equilibrium water loading of the zeolite at the specified steady-state conditions.

From the equilibrium water loading w of the zeolite at different adsorber temperatures T_{ads} and evaporator temperatures T_{evap}, we calculate the adsorbate volume W

$$W(w, T_{\mathrm{ads}}) = \frac{w}{\varrho_{\mathrm{ad}}(T_{\mathrm{ads}})} \tag{A.1}$$

and the adsorption potential A (cf. Equation (7.2))

$$A(T_{\text{ads}}, T_{\text{evap}}) = RT_{\text{ads}} \ln \left(\frac{p_{\text{s}}(T_{\text{ads}})}{p_{\text{s}}(T_{\text{evap}})} \right) \tag{A.2}$$

for the characteristic function of the zeolite. To obtain a complete characteristic function, we repeated the measurement procedure several times for different temperatures. The resulting data is shown as red dots in Figure A.1. With this data, we fit the parameters of Equation (7.1) to describe the characteristic function for the zeolite (cf. blue line in Figure A.1).

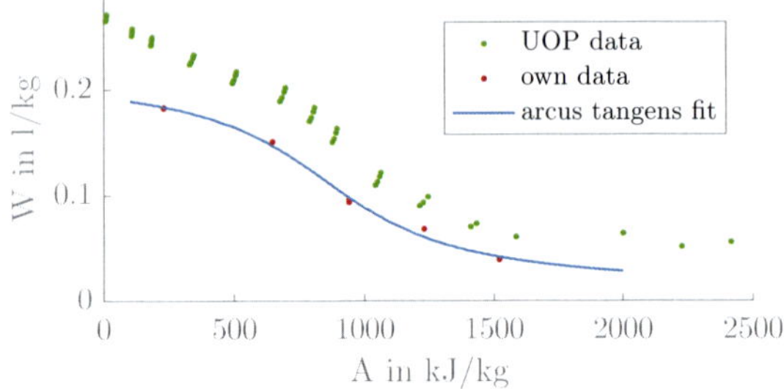

Figure A.1: Characteristic function for zeolite 13 X according to manufacturer data (UOP) and own measurements - reduced capacity visible.

A comparison of our measurements to manufacturer data (cf. green dots in Figure A.1) reveals a capacity reduction to approximately 75 %. A similar reduction of the adsorption capacity has been reported for binderless zeolite 13 X by Jänchen et al. [162].

A.2 Heat-transfer-fluid circulation systems

In the experimental setup (cf. Chapter 5), the heat exchangers of the adsorption TES unit are connected to circulation systems with water and a thermal oil. Relevant data on pumps, flow meters and the thermal oil are given in the following:

Water pumps and volume flow meters

The water in the circulation system is pumped with two centrifugal pumps of type Wilo-Economy MHI 202 by Wilo SE from Hilden, Germany. The volume flow rate of water is measured with vortex flow meters of type 230 by Huba Control AG from Würenlos, Switzerland. The flow control adjusts the volume flow rate of water to 9 L/min.

Heat-transfer oil

The heat-transfer oil in the circulation system of our experimental setup is XW15 by ADDINOL Lube Oil GmbH - High-Performance Lubricants from Leuna, Germany. This synthetic fluid is based on alkylbenzenes, and is applicable up to 300 °C. The density ϱ, heat capacity c_p and dynamic viscosity η are given with the oil temperature T in °C:

$$\varrho_\mathrm{oil} = 900.6\,\mathrm{kg/m^3} - 0.6323\,\mathrm{kg/m^3} \cdot T/{}^\circ\mathrm{C} - 0.0002\,\mathrm{kg/m^3} \cdot T^2/{}^\circ\mathrm{C}^2 \tag{A.3}$$

$$c_\mathrm{p,oil} = 1.392\,\mathrm{kJ/(kg\,K)} + 0.0102\,\mathrm{kJ/(kg\,K)} \cdot T/{}^\circ\mathrm{C} - 1.969 \times 10^{-5}\,\mathrm{kJ/(kg\,K)} \cdot T^2/{}^\circ\mathrm{C}^2 \tag{A.4}$$

$$\eta_\mathrm{oil} = 43\,220\,\mathrm{Pa\,s} \cdot (T/{}^\circ\mathrm{C})^{-2.061}. \tag{A.5}$$

These equations are valid from 50 – 250 °C.

Oil pump and mass flow meter

The oil in the circulation system is pumped with a centrifugal pump of type CY-4281-MK-TOE Q90 by Speck Pumpen Systemtechnik GmbH from Roth, Germany. The mass flow rate of the oil is measured with a Coriolis mass flow meter of type TMU015 by Heinrichs Messtechnik GmbH from Cologne, Germany. The flow control adjusts the mass flow rate of the thermal oil to mostly 300 kg/h.

Determination of heat losses of the oil circulation system

While at temperatures above its environment, the oil circulation system suffers from heat losses. These losses have to be taken into account when solving the energy balance around the adsorber, since the temperature measurement positions of adsorber inlet and outlet are each 0.55 m from the adsorber.

For the determination of the heat losses of the complete oil circulation system, the adsorber is bypassed and the circulation system is heated to a constant temperature until a steady state is reached. In steady state, the heat losses of the oil circulation system equal the heat input $\dot{Q}_{\text{steady-state, in}} = \dot{Q}_{\text{steady-state, loss}}$. The heat losses of the oil circulation system are calculated considering the complete length of the circulation system (6 m). A corresponding heat-loss coefficient $(UA)_{\text{circulation}}$ can be calculated by

$$(UA)_{\text{circulation}} = \frac{\dot{Q}_{\text{steady-state, loss}}}{T_{\text{oil,ave}} - T_{\text{env}}}, \tag{A.6}$$

with the ambient temperature T_{env} and an average temperature $T_{\text{oil,ave}}$ of the thermal oil in the circulation system.

In the calculations in Section 6.2.2 and Section 7.2.1, the heat losses of the oil circulation system are included with the coefficient $(UA)_{\text{pipe-loss}}$, which is related to the length of the oil pipe between the adsorber and the measurement positions of adsorber inlet $T_{\text{A,in}}$ and outlet $T_{\text{A,out}}$. We obtain this heat-loss coefficient $(UA)_{\text{pipe-loss}}$ by

$$(UA)_{\text{pipe-loss}} = 0.55\,\text{m}\frac{(UA)_{\text{circulation}}}{6\,\text{m}}. \tag{A.7}$$

Figure A.2 shows the measured heat losses of the oil pipes between the adsorber and the measurement positions of adsorber inlet $T_{\text{A,in}}$ and outlet $T_{\text{A,out}}$ as a function of the oil temperature T_{oil}.

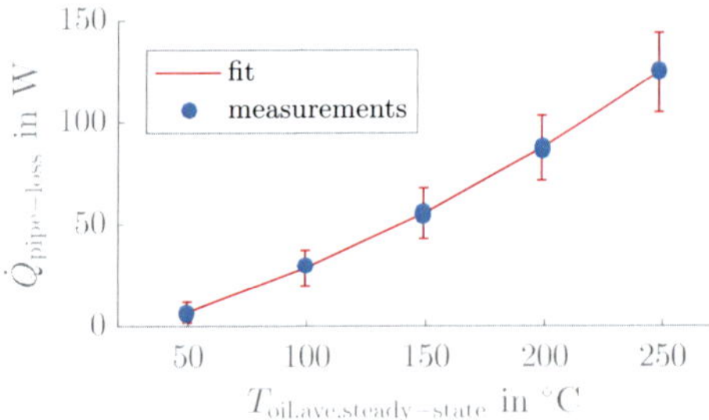

Figure A.2: Heat losses of the oil circulation system. The error bars show the measurement uncertainty of the heat losses.

In Chapter 6, the efficiencies are calculated for the adsorber only, i.e. further losses of the oil circulation system are neglected. For investigations of process integration, the heat losses of the pipes connecting the storage unit with the heat sources and sinks have to be taken into account. These losses can be significant for long distances between the storage unit and the corresponding process setup. Thus, a good insulation of the circulation system is necessary.

B Measurement Uncertainty

This appendix provides details on the measurement uncertainties of the two experimental setups that are the basis for the thorough evaluation of our adsorption thermal energy storage (TES) unit. We determine the expanded measurement uncertainty according to the *Guide to the Expression of Uncertainty in Measurement* (GUM) [132] with a coverage probability of 95 %.

For any physical quantity $Y = f(x_i)$ which is calculated from measured values x, we calculate the combined uncertainty u_Y according to the law of propagation of uncertainty for uncorrelated values:

$$u_Y = \sqrt{\sum_{\mathrm{i}} \left(\frac{\partial f}{\partial x_{\mathrm{i}}}\right)^2 {u_{x_{\mathrm{i}}}}^2}, \tag{B.1}$$

where $\frac{\partial f}{\partial x_{\mathrm{i}}}$ is the partial derivative of f with respect to x_{i}.

For the measurement equipment, we refer to the manufacturers' specifications of the measurement uncertainty u_x and assume a rectangular probability distribution, as proposed by GUM [132]. The expanded measurement uncertainty $U_x^{95\%}$ is then calculated by

$$U_x^{95\%} = \frac{u_x}{\sqrt{3}} 1.65 = 0.95\, u_x. \tag{B.2}$$

The measurement uncertainty of Pt100 resistance thermometers can be calculated by

$$u_{\mathrm{Pt100}} = 0.15\,\mathrm{K} + 0.002 \cdot |T - 273.15\,\mathrm{K}| \tag{B.3}$$

from DIN IEC 60751 [163] for class A and four-wire configuration. The uncertainty of the thermocouples K-type u_{TC} is 1.5 K according to DIN IEC 60584-2 [164] for class 1.

B.1 Experimental setup for brewery investigations

This section contains details on the measurement uncertainty of the experimental setup shown in Figure 4.2. The measurement uncertainty of the adsorber heat flow rate $U^{95\%}_{\dot{Q}_{\mathrm{Ads}}}$ is determined according to the propagation of uncertainty [132], cf. Equation (B.1), with the following assumptions: the relative measurement deviation of the used flow meters is 1 % of the measured value according to the data sheet of the manufacturer. The measurement deviation of the used thermometers can be calculated according to Equation (B.3). Further uncertainties of fluid properties (density ϱ and heat capacity c_{p}) are neglected.

The absolute measurement uncertainty of the heat flow rates $U^{95\%}_{\dot{Q}}$ of the experimental setup (cf. Figure 4.2) is in the range of 300–800 W. The relative measurement uncertainty for the transferred heat in the adsorber $U^{95\%}_{Q\mathrm{ads,rel}}$ is calculated by:

$$U^{95\%}_{Q\mathrm{ads,rel}} = \frac{\int U^{95\%}_{\dot{Q}_{\mathrm{ads}}}(t)\,\mathrm{d}t}{\int \dot{Q}_{\mathrm{ads}}(t)\,\mathrm{d}t} \tag{B.4}$$

with the time dependent heat flow rate $\dot{Q}_{\mathrm{ads}}(t)$. Equation (B.4) assumes total correlation between the heat flow rate measurements [132]. The relative measurement uncertainty of the heat flow rates during desorption $U^{95\%}_{\dot{Q}_{\mathrm{des,rel}}}$ varies between 29.4 and 62 %, resulting in a relative measurement uncertainty of the transferred heat during desorption of $U^{95\%}_{Q\mathrm{des,rel}} =$ 44.1 %. During adsorption, the relative measurement uncertainty of the heat flow rates $U^{95\%}_{\dot{Q}_{\mathrm{ads,rel}}}$ varies between 4.9 and 17.6 %. The measurement uncertainty of the transferred heat during adsorption is $U^{95\%}_{Q\mathrm{ads,rel}} = 7.3\,\%$.

The relative uncertainty is larger for lower heat flow rates, because the measured temperature difference over the adsorber heat exchanger reached values close to the measurement uncertainty of the thermometers. Unfortunately, it was not possible to reduce this uncertainty, since the volume flow could not be reduced. Nevertheless, for similar experimental setups, the accuracy is much better even for lower heat flow rates (cf. Lanzerath et al. [82]). Thus, based on our experience we expect the quality of the trends of the measured adsorber heat flow rates to be acceptable.

Unfortunately, the heat flow rates to the evaporator cannot be properly resolved, because a large volume flow rate leads to temperature differences over the evaporator which are below the uncertainty of the thermometers.

B.2 Experimental setup for adsorption thermal energy storage units

This section provides details on the determination of the expanded measurement uncertainty $U^{95\%}$ of the experimental setup described in Chapter 5. The expanded uncertainties of the measurement devices and the calibrated temperature differences are given in the following paragraphs. These uncertainties combine to the expanded measurement uncertainty of the steady-state heat flow rate $U^{95\%}_{\dot{Q}_{\text{steady-state, input}}}$ (cf. Table 6.1) according to uncertainty propagation (cf. Equation (B.1)). We again neglect uncertainties of the fluid properties density ϱ and heat capacity c_p. For all values subsequently derived from the measured heat flow rates, expanded uncertainties are also calculated according to uncertainty propagation [132].

Fluid flow measurements

The measurement uncertainty of the Coriolis Mass Flow Meter TMU is given by Equation (B.5) from the data sheet of the manufacturer (Heinrichs Messtechnik GmbH, Cologne, Germany)

$$u_{\dot{m}} = 1.2\,\text{kg/h} + 0.1\,\dot{m}/100 \tag{B.5}$$

with the measured value of the mass flow rate $\dot{m}$ in kg/h.

Measurements of temperature differences

For the temperature difference ΔT between the adsorber inlet and outlet, a statistical deviation was calculated from six calibration measurements. This deviation is multiplied with the student's t-distribution factor $k_t = 2.571$ [165] to obtain the expanded uncertainty of $U^{95\%}_{\Delta T} = 0.057\,\text{K}$.

The calculation of the difference of the enthalpy flow rate $\Delta\dot{H}_\text{oil}$ (cf. Equation (6.2)) includes a correction with measured heat losses of the oil pipes between the adsorber and the measurement positions of adsorber in- and outlet. These heat losses have been estimated experimentally (cf. Section A.2) and the resulting statistical deviation from 3 measurements is multiplied with the student's t-distribution factor $k_t = 12.71$ [165] to obtain the expanded uncertainty of $U^{95\%}_\text{pipe-loss}$. A possible systematic error due to the correction of $\Delta\dot{H}_\text{oil}$ cannot be determined. Hence, we compared our findings calculated with and without this correction and confirmed that results are qualitatively the same.

Pressure measurement

The uncertainty of the pressure sensor is $u_p = 8\,\text{mbar}$. This uncertainty leads to inaccuracies during the determination of the heat-transfer coefficients of evaporation and condensation in Section 7.2.1.

Figure B.1 shows the evaporator heat-transfer coefficient $(UA)_{\text{E,HX}}$ (blue line) resulting from Equation (7.17) and the logarithmic temperature difference ΔT_{ln} (red line) throughout the evaporator heat exchanger tube, calculated with the saturation temperature T_{s} of the measured pressure in the evaporator p_{E}. To show the influence of the uncertain pressure

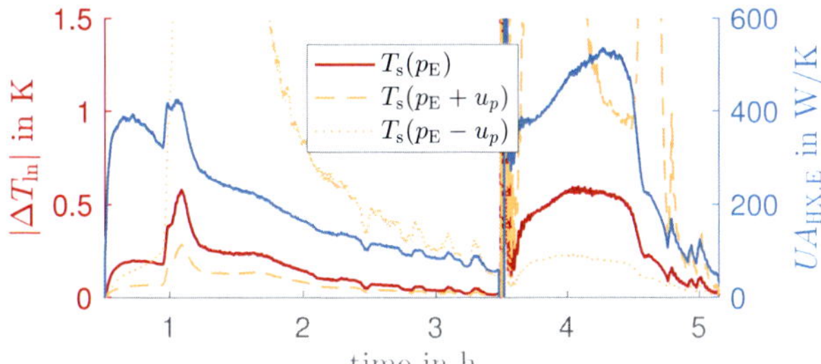

Figure B.1: Heat-transfer coefficient $(UA)_{\text{E,HX}}$ resulting from Equation (7.17) and logarithmic temperature difference ΔT_{ln} throughout the evaporator heat exchanger tube, calculated with the saturation temperature T_{s} of the measured pressure in the evaporator p_{E} (red line) and an upper and lower limit with the uncertainty of the pressure sensor u_p (yellow lines).

measurement on the temperature, Figure B.1 further contains the logarithmic temperature difference ΔT_{ln} for an upper and lower limit of the pressure p_{E} with the uncertainty of the pressure sensor u_p. The influence on the temperature difference leads to a deviation of more than 50 %.

The slope of the temperature difference mainly determines the slope of the heat-transfer coefficient. Thus, uncertainties in the determination of the temperatures have a large impact on the heat-transfer coefficient in the evaporator. The low accuracy of the pressure sensor is propagated to the temperature of the adsorptive water in the evaporator and thus directly influences the accuracy of the heat-transfer coefficient.

C Auxiliary Model Data

This appendix provides auxiliary data of the models, which are used in this thesis. Moreover, all simulation results of the validation measurements are given in Section C.2.

C.1 Model parameters

The model parameters used during the simulation study in Chapter 4 can be found in Table C.1.

Table C.1: Parameter values of the adsorption TES model.

Parameter	Value
m_{ads}	20 kg
c_{ads}	1000 J/(K kg)
r_{ads}	0.7 mm
V_{E}	0.04 m^3
$C_{\mathrm{A,cas}}$	5700 J/K
c_{steel}	500 J/(K kg)
$m_{\mathrm{E,met}}$	3.06 kg
$m_{\mathrm{A,lam}}$	4.76 kg
T_{env}	298.15 K

C.2 Model validation

This section contains figures of the temperature and heat flow rates from all remaining validation measurements (cf. Table 7.4) which are not shown in Chapter 7. The corresponding simulation accuracies are shown in Figure 7.11 and in Figure 7.12.

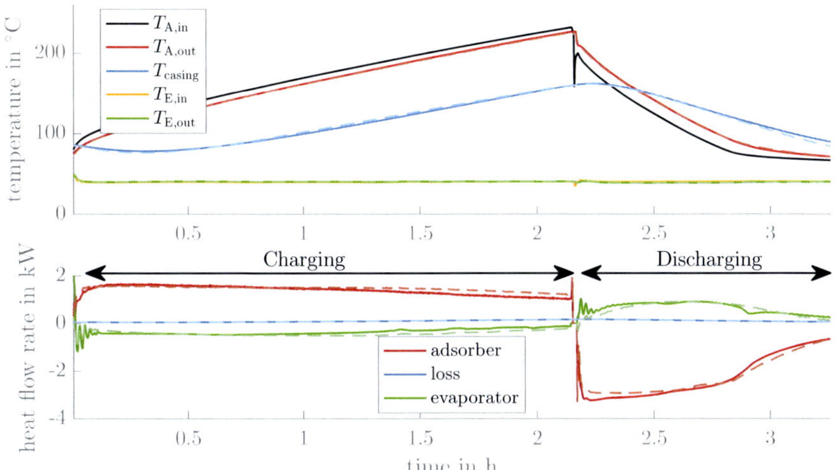

Figure C.1: Validation measurement with direct discharge, $T_{des} = 225\,°C$ and $T_{evap} = 40\,°C$. Top: temperatures of adsorber and evaporator. Bottom: heat flow rate of adsorber, evaporator and losses to the environment. Dashed lines show the corresponding simulation results.

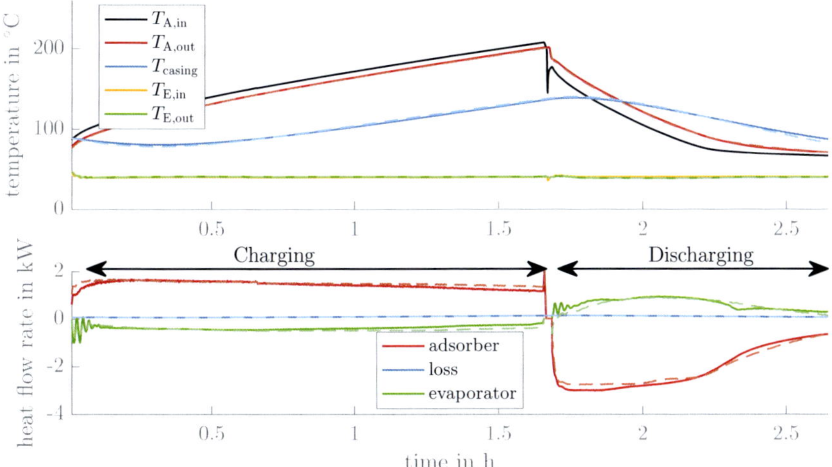

Figure C.2: Validation measurement with direct discharge, $T_{des} = 200\,°C$ and $T_{evap} = 40\,°C$. Top: temperatures of adsorber and evaporator. Bottom: heat flow rate of adsorber, evaporator and losses to the environment. Dashed lines show the corresponding simulation results.

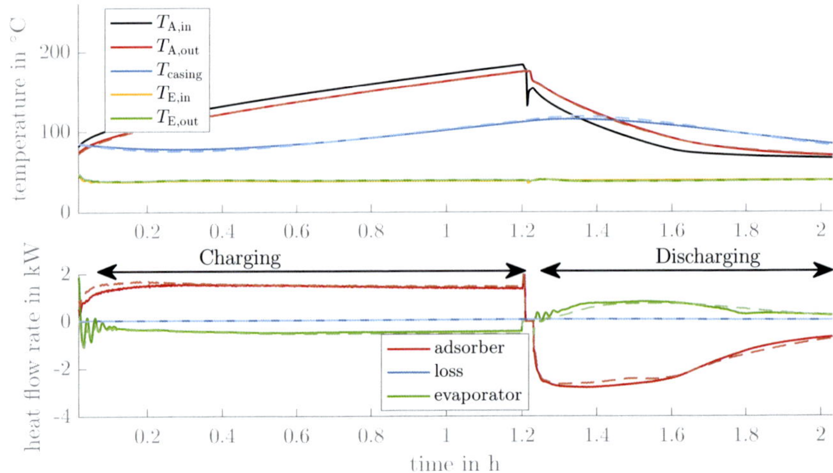

Figure C.3: Validation measurement with direct discharge, $T_{des} = 175\,°C$ and $T_{evap} = 40\,°C$. Top: temperatures of adsorber and evaporator. Bottom: heat flow rate of adsorber, evaporator and losses to the environment. Dashed lines show the corresponding simulation results.

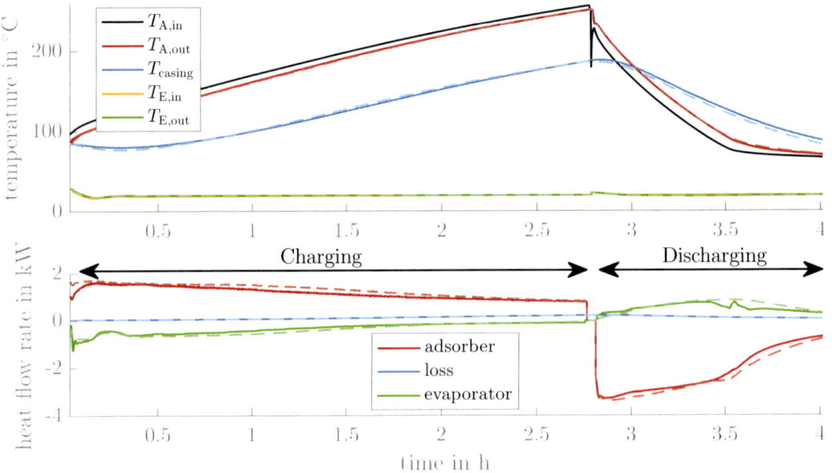

Figure C.4: Validation measurement with direct discharge, $T_{des} = 250\,°C$ and $T_{evap} = 20\,°C$. Top: temperatures of adsorber and evaporator. Bottom: heat flow rate of adsorber, evaporator and losses to the environment. Dashed lines show the corresponding simulation results.

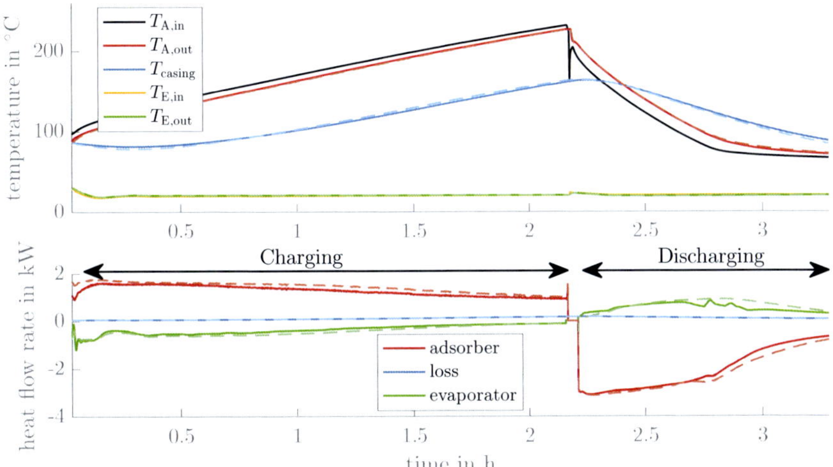

Figure C.5: Validation measurement with direct discharge, $T_{des} = 225\,°C$ and $T_{evap} = 20\,°C$. Top: temperatures of adsorber and evaporator. Bottom: heat flow rate of adsorber, evaporator and losses to the environment. Dashed lines show the corresponding simulation results.

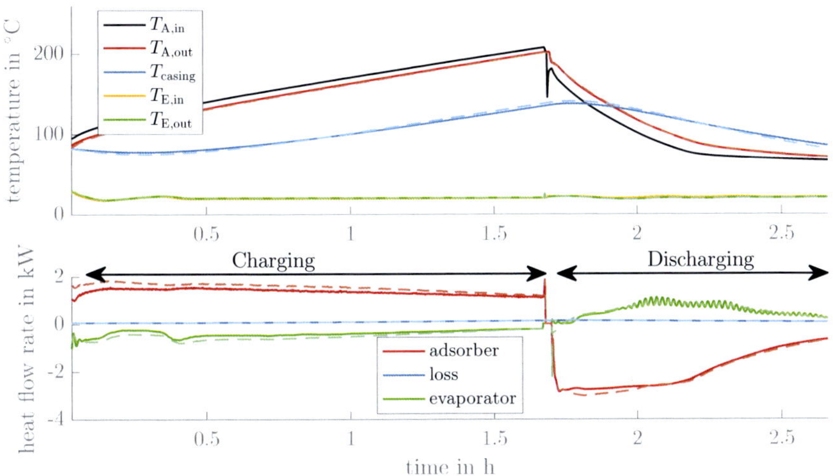

Figure C.6: Validation measurement with direct discharge, $T_{des} = 200\,°C$ and $T_{evap} = 20\,°C$. Top: temperatures of adsorber and evaporator. Bottom: heat flow rate of adsorber, evaporator and losses to the environment. Dashed lines show the corresponding simulation results.

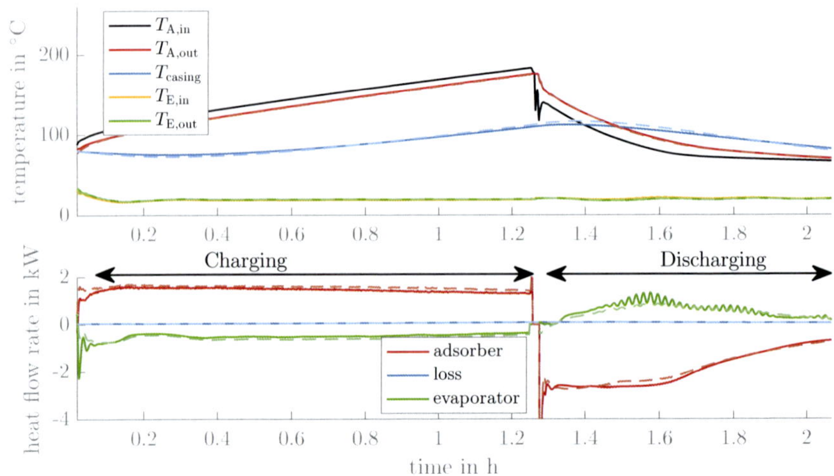

Figure C.7: Validation measurement with direct discharge, $T_{des} = 175\,°C$ and $T_{evap} = 20\,°C$. Top: temperatures of adsorber and evaporator. Bottom: heat flow rate of adsorber, evaporator and losses to the environment. Dashed lines show the corresponding simulation results.

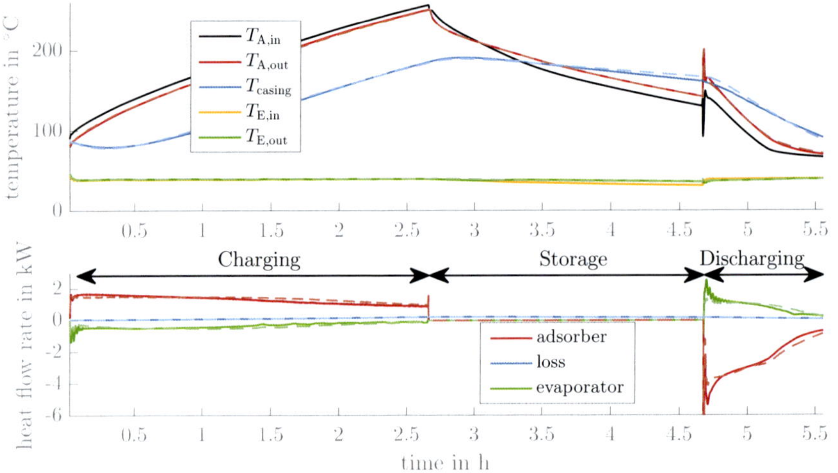

Figure C.8: Validation measurement with 2 h storage time, $T_{des} = 250\,°C$ and $T_{evap} = 40\,°C$. Top: temperatures of adsorber and evaporator. Bottom: heat flow rate of adsorber, evaporator and losses to the environment. Dashed lines show the corresponding simulation results.

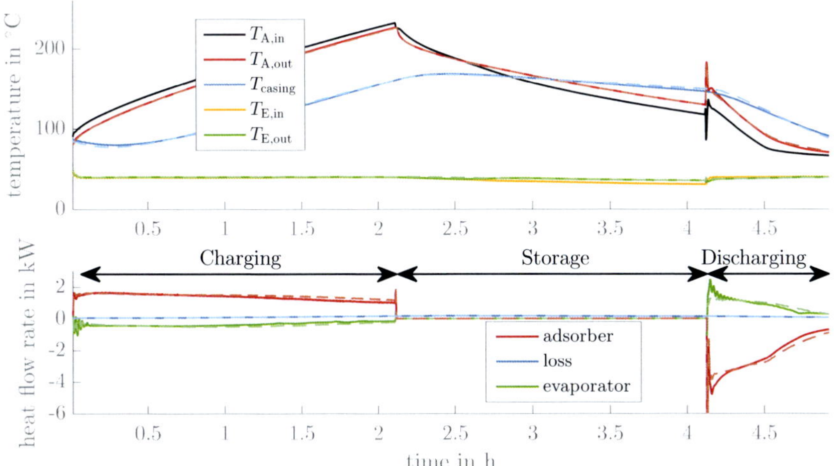

Figure C.9: Validation measurement with 2 h storage time, $T_{des} = 225\,°C$ and $T_{evap} = 40\,°C$. Top: temperatures of adsorber and evaporator. Bottom: heat flow rate of adsorber, evaporator and losses to the environment. Dashed lines show the corresponding simulation results.

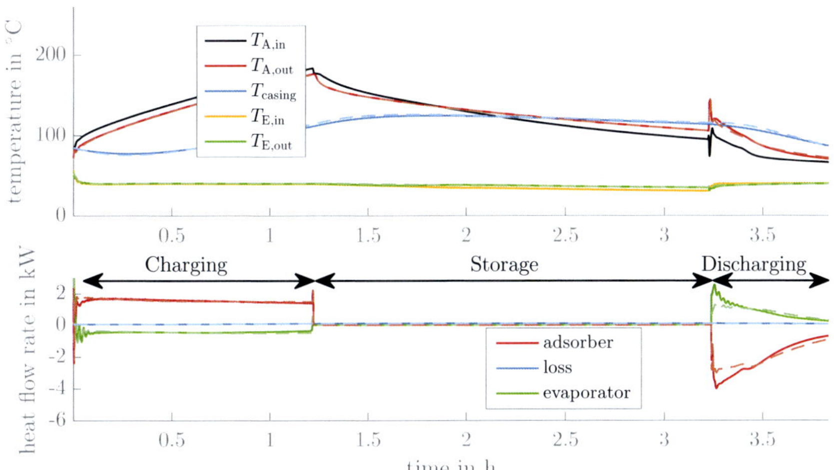

Figure C.10: Validation measurement with 2 h storage time, $T_{des} = 175\,°C$ and $T_{evap} = 40\,°C$. Top: temperatures of adsorber and evaporator. Bottom: heat flow rate of adsorber, evaporator and losses to the environment. Dashed lines show the corresponding simulation results.

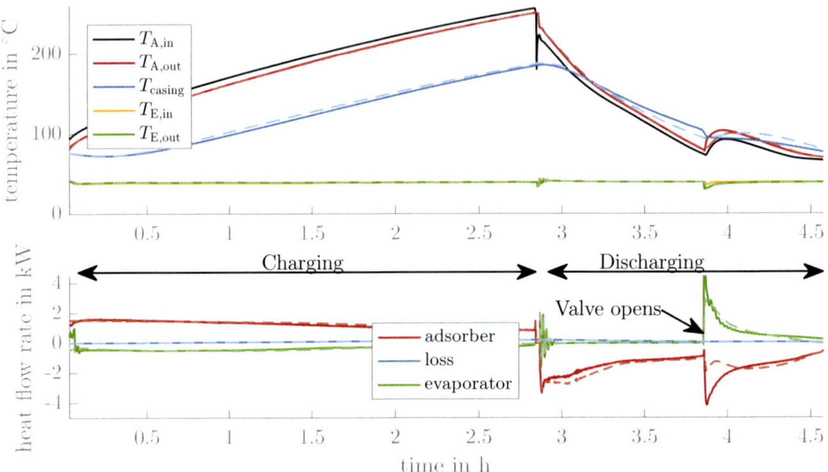

Figure C.11: Validation measurement with closed valve at beginning of adsorption, $T_{des} = 250\,°C$ and $T_{evap} = 40\,°C$. Top: temperatures of adsorber and evaporator. Bottom: heat flow rate of adsorber, evaporator and losses to the environment. Dashed lines show the corresponding simulation results.

D List of Publications

Journal paper

U. Bau, P. Hoseinpoori, S. Graf, H. Schreiber, F. Lanzerath, C. Kirches, and A. Bardow. "Dynamic optimisation of adsorber-bed designs ensuring optimal control". *Applied Thermal Engineering* 125 (2017), pp. 1565–1576.

H. Schreiber, F. Lanzerath, C. Reinert, C. Grüntgens, and A. Bardow. "Heat lost or stored: Experimental analysis of adsorption thermal energy storage". *Applied Thermal Engineering* 106 (2016), pp. 981–991.

F. Lanzerath, J. Seiler, M. Erdogan, H. Schreiber, M. Steinhilber, and A. Bardow. "The impact of filling level resolved: Capillary-assisted evaporation of water for adsorption heat pumps". *Applied Thermal Engineering* 102 (2016), pp. 513–519.

H. Schreiber, S. Graf, F. Lanzerath, and A. Bardow. "Adsorption thermal energy storage for cogeneration in industrial batch processes: Experiment, dynamic modeling and system analysis". *Applied Thermal Engineering* 89 (2015), pp. 485–493.

H. Schreiber, B. Klitzing, F. Lanzerath, A. Gebhard, and A. Bardow. "Integrating Cogeneration and Heat Storage for Energy-Efficient Industrial Batch Processing". *Euro Heat & Power* 10.1 (2013), pp. 42–45.

M. Roeb, J.-P. Säck, P. Rietbrock, C. Prahl, H. Schreiber, et al.. "Test operation of a 100 kW pilot plant for solar hydrogen production from water on a solar tower". *Solar Energy* 85.4 (2011), pp. 634–644.

Conference contributions

H. Schreiber, F. Lanzerath, und A. Bardow, "Predicting Performance of Adsorption Thermal Energy Storage: From Experiments to Validated Dynamic Models". Poster, *4th Sustainable Thermal Energy Management International Conference SusTEM2017*. Alkmaar, Netherlands.

H. Schreiber, U. Bau, F. Lanzerath, und A. Bardow, "Thermal Energy Storage Using Adsorption: Is it Really Lossless?". Poster, *ProcessNet-Jahrestagung 2016*. Aachen, Germany and *Chemie Ingenieur Technik* 88.9 (2016) p.1274.

H. Schreiber, U. Bau, F. Lanzerath, und A. Bardow, "Heat loss in adsorption thermal energy storage: modeling and experimental validation". Oral presentation, *Heat Powered Cycles 2016*. Nottingham, United Kingdom.

U. Bau, P. Hoseinpoori, S. Graf, H. Schreiber, F. Lanzerath, und A. Bardow, "Rigorous assessment of adsorber-bed designs using dynamic optimization". Oral presentation, *Heat Powered Cycles 2016*. Nottingham, United Kingdom.

U. Bau, H. Schreiber, F. Lanzerath, und A. Bardow. "Hybride Klimatisierung für Elektrofahrzeuge mit Wärmepumpe und offenem Sorptionssystem" (in German). Oral presentation, *Deutsche Kälte- und Klimatagung 2016*. Kassel, Germany.

U. Bau, H. Schreiber, F. Lanzerath, und A. Bardow. "Hybride Klimatisierung für Elektrofahrzeuge: Kombination von Wärmepumpe und offenem Sorptionssystem" (in German). Oral presentation, *Wärmemanagement des Kraftfahrzeugs X 2016*. Potsdam, Germany.

U. Bau, H. Schreiber, F. Lanzerath, und A. Bardow. "Adsorption-based air-conditioning for battery-driven electric busses", Oral presentation, *24th IIR International Congress of Refrigeration 2015*. Tokio, Japan.

F. Lanzerath, H. Schreiber, U. Bau, und A. Bardow. "Adsorption-based Air-Conditioning for Battery-driven Electric Busses", Poster, *Sorption Friends 2015*. Milazzo, Italy.

U. Bau, F. Lanzerath, M. Gräber, H. Schreiber, N. Thielen und A. Bardow, "Adsorption energy systems library - Modeling adsorption based chillers, heat pumps, thermal storages and desiccant systems". Oral presentation, *10th International Modelica Conference 2014*. Linköping, Sweden.

H. Schreiber, U. Bau, und A. Bardow, "Sorptionsgestützte Klimatisierung von Elektrobussen" (in German). Oral presentation, *Thermodynamik-Kolloquium 2014*. Stuttgart, Germany.

H. Schreiber, S. Graf, F. Lanzerath, and A. Bardow. "Adsorption heat storage for combined heat and power units in industrial batch processes". Oral presentation, *International Sorption Heat Pump Conference 2014*. College Park, Md, USA.

F. Lanzerath, J. Seiler, M. Erdogan, H. Schreiber, M. Steinhilber, and A. Bardow. "Combination of finned tubes and thermal coating for high performance water evaporation in adsorption heat pumps". Oral presentation, *International Sorption Heat Pump Conference 2014*. College Park, Md, USA.

S. Graf, H. Schreiber, and A. Bardow. "Model-based evaluation of adsorption heat storage for combined heat and power". Poster, *26. Deutsche Zeolith-Tagung 2014*, Paderborn, Germany.

S. Graf, H. Schreiber, and A. Bardow. "Modellgestützte Bewertung von Adsorptionswärmespeichern für industrielle Kraft-Wärme-Kopplung" (in German). Poster, *Thermodynamik-Kolloquium 2013*, Hamburg, Germany.

F. Lanzerath, H. Schreiber, M. Erdogan, and A. Bardow. "Sorption Systems Engineering". Poster, *Heat Powered Cycles 2012*, Alkmaar, The Netherlands.

H. Schreiber, B. Klitzing, F. Lanzerath, A. Gebhard, and A. Bardow. "Integrating cogeneration and heat storage for an energy-efficient industrial batch process". Poster, *7th International Renewable Energy Storage Conference and Exhibition 2012*, Berlin, Germany.

H. Schreiber, B. Klitzing, and A. Bardow. "Zeolithspeicher zur Lastverschiebung für die energieeffiziente Wärmeversorgung einer Brauerei" (in German). Poster, *Thermodynamik-Kolloquium 2011*, Frankfurt, Germany.

B. Klitzing, H. Schreiber, F. Lanzerath, and A. Bardow. "Modellgestützte Analyse der Adsorption in mikroporösen Materialien" (in German). Oral presentation, *Jahrestreffen des Fachausschusses Adsorption 2011*, Würzburg, Germany.

H. Schreiber, B. Klitzing. "On model-based determination of physical properties for adsorption working pairs". Poster, *Cape Forum - Computer Aided Process Engineering 2010*, Aachen, Germany.

E Student theses supervised during this work

Christiane Reinert and Sophia Schröer (2015). Experimentelle und simulationsgestützte Bewertung eines thermischen Adsorptionsspeichers für Kraft-Wärme-Kopplungs-Prozesse (in German). Project thesis, RWTH Aachen University

Christoph Grüntgens (2015). Experimentelle und modellgestützte Untersuchung von Wärmeverlusten an einem thermischen Adsorptionsspeicher (in German). Bachelor thesis, RWTH Aachen University

Julia Moll (2015). Auslegung und Implementierung der Temperatur- und Leistungsregelung eines flexiblen Prüfstands für thermische Adsorptionsspeicher (in German). Master thesis, RWTH Aachen University

Stephan Merkelbach (2013). Konzeption eines flexiblen Prüfstands für Adsorptionswärmespeicher (in German). Master thesis, RWTH Aachen University

Thomas Michael Bauer (2013). Wärme- und Stofftransport in einem Adsorptionsspeicher: FEM Simulation und experimentelle Validierung (in German). Diploma thesis, RWTH Aachen University

Stefan Graf (2013). Modellgestützte Bewertung von Adsorptionswärmespeichern für industrielle Kraft-Wärme-Kopplung (in German). Master thesis, RWTH Aachen University

Philipp Zur-Lienen (2013). Konzept und Modellierung eines stromgeführten BHKW mit Adsorptionswärmespeicher (in German). Bachelor thesis, RWTH Aachen University

Andreas Gebhardt (2013). Experimentelle Untersuchung des Einflusses der Niedertemperaturwärme auf die Speicherkapazität einer Adsorptions-Anlage (in German). Semester thesis, RWTH Aachen University

Lorenz Julian Frigge (2012). Experimentelle Untersuchung und Simulation der elektrischen Desorption eines Zeolith-Wasser-Wärmespeichers (in German). Semester thesis, RWTH Aachen University

Thomas Michael Bauer (2012). Theoretische Betrachtung, Aufbau und Messung einer thermisch betriebenen Wärmepumpe, sowie die theoretische Betrachtung alternativer Konzepte (in German). Semester thesis at Bosch Thermotechnik GmbH, RWTH Aachen University

Holger Schmeing (2012). Experimental Determination of Potential Energy Savings by Deployment of Phase-Change Material in a Commercial Fridge/Freezer. Project thesis at University of California, Davis, Ca, USA, RWTH Aachen University

Jan Henning Jockenhöfer (2012). Experimentelle und theoretische Untersuchungen eines Festbettreaktors zur thermochemischen Speicherung niederkalorischer Wärme (in German). Diploma thesis at DLR Stuttgart, German Aerospace Center, RWTH Aachen University

Kerstin Bernicke (2012). Bestimmung der Wärmekapazität von Zeolith (in German). Semester thesis, RWTH Aachen University

Bibliography

[1] R. Pachauri and L. Meyer. *Climate Change 2014: Sythesis Report.* Contribution of Working Groups I, II and III to the Fifth Assessment Report. Geneva, Switzerland: Intergovernmental Panel in Climate Change (IPCC), 2014.

[2] H. Kondziella and T. Bruckner. "Flexibility requirements of renewable energy based electricity systems – a review of research results and methodologies". *Renewable and Sustainable Energy Reviews* 53 (2016), pp. 10–22.

[3] A. B. Gallo, J. R. Simões-Moreira, H. Costa, M. M. Santos, and E. Moutinho dos Santos. "Energy storage in the energy transition context: A technology review". *Renewable and Sustainable Energy Reviews* 65 (2016), pp. 800–822.

[4] M. C. Lott and S.-I. Kim. *Technology Roadmap Energy Storage.* Tech. rep. International Energy Agency, 2014.

[5] *Ausgewählte Grafiken zu Energiegewinnung und Energieverbrauch: Energieverbrauch nach Anwendungsbereichen in Deutschland 2014 (in German).* online, accessed 21/10/2016. Bundesministerium für Wirtschaft und Energie. 2014. URL: `www.bmwi.de/DE/Themen/Energie/Energiedaten-und-analysen/Energiedaten/energiegewinnung-energieverbrauch.html`.

[6] L. Hyman. *Sustainable Thermal storage Systems Planning, Desing, and Operations.* New York, United States of America: McGraw-Hill, 2011.

[7] S. Kalaiselvam and R. Parameshwaran, eds. *Thermal Energy Storage Technologies for Sustainability - Systems Design, Assessment and Applications.* Boston, United States of America: Academic Press, 2014.

[8] N. Yu, R. Wang, and L. Wang. "Sorption thermal storage for solar energy". *Progress in Energy and Combustion Science* 39 (2013), pp. 489–514.

[9] J.-C. Hadorn, B. Lehmann, R. Weber, T. Letz, W. Streicher, C. Bales, M. Haller, P. Vogelsanger, J. Schulz, S. Citherlet, M. Yamaha, L. F. Cabeza, A. Heinz, P. Gantenbein, A. Hauer, D. Jaehnig, H. Kerskes, H.-M. Henning, T. Núñez, K. Visscher, E. Lävemann, M. Peltzer, and J. V. Berkel. *Thermal Energy Storage for*

Solar and Low Energy Buildings - State of the Art. Ed. by J.-C. Hadorn. Lleida, Spain: Edicions de la Universitat de Lleida, 2005.

[10] M. Seiler, A. Kühn, F. Ziegler, and X. Wang. "Sustainable Cooling Strategies Using New Chemical System Solutions". *Industrial & Engineering Chemistry Research* 52 (2013), pp. 16519–16546.

[11] J. Kariya, J. Ryu, and Y. Kato. "Development of thermal storage material using vermiculite and calcium hydroxide". *Applied Thermal Engineering* 94 (2016), pp. 186–192.

[12] Y. I. Aristov. "Challenging offers of material science for adsorption heat transformation: A review". *Applied Thermal Engineering* 50.2 (2013). Combined Special Issues: ECP 2011 and IMPRES 2010, pp. 1610–1618.

[13] K. Korhammer, M.-M. Druske, A. Fopah-Lele, H. U. Rammelberg, N. Wegscheider, O. Opel, T. Osterland, and W. Ruck. "Sorption and thermal characterization of composite materials based on chlorides for thermal energy storage". *Applied Energy* 162 (2016), pp. 1462–1472.

[14] W. van Helden and M. Rommel. "Compact Thermal Energy Storage: Material Development for System Integration". *Proceedings of the solar heating and cooling conference (SHC).* Istanbul, Turkey, 2015.

[15] J. Jänchen, T. H. Herzog, K. Gleichmann, B. Unger, A. Brandt, G. Fischer, and H. Richter. "Performance of an open thermal adsorption storage system with Linde type A zeolites: Beads versus honeycombs". *Microporous and Mesoporous Materials* 207 (2015), pp. 179–184.

[16] D. Aydin, S. P. Casey, and S. Riffat. "The latest advancements on thermochemical heat storage systems". *Renewable and Sustainable Energy Reviews* 41 (2015), pp. 356–367.

[17] M. M. Heravi, M. Tajbakhsh, A. N. Ahmadi, and B. Mohajerani. "Zeolites. Efficient and Eco-friendly Catalysts for the Synthesis of Benzimidazoles". *Chemical Monthly* 137.2 (2006), pp. 175–179.

[18] H. Karge and J. Weitkamp, eds. *Zeolites as Catalysts, Sorbents and Detergent Builders.* 1st ed. Vol. 46. Amsterdam, The Netherlands: Elsevier Science, 1989.

[19] A. Hauer. "Beurteilung fester Adsorbentien in offenen Sorptionssystemen für energetische Anwendungen (in German)". Dissertation. Berlin, Germany: Technische Universität Berlin, 2002.

[20] A. Lass-Seyoum, M. Blicker, D. Borozdenko, T. Friedrich, and T. Langhof. "Transfer of laboratory results on closed sorption thermo-chemical energy storage to a large-scale technical system". *Energy Procedia* 30 (2012). Proceedings of the 1st International Conference on Solar Heating and Cooling for Buildings and Industry (SHC 2012), pp. 310–320.

[21] Y. Lu, R. Wang, M. Zhang, and S. Jiangzhou. "Adsorption cold storage system with zeolite-water working pair used for locomotive air conditioning". *Energy Conversion and Management* 44.10 (2003), pp. 1733–1743.

[22] Y. Criado, M. Alonso, J. Abanades, and Z. Anxionnaz-Minvielle. "Conceptual process design of a $CaO/Ca(OH)_2$ thermochemical energy storage system using fluidized bed reactors". *Applied Thermal Engineering* 73.1 (2014), pp. 1087–1094.

[23] A. Solé, I. Martorell, and L. F. Cabeza. "State of the art on gas-solid thermochemical energy storage systems and reactors for building applications". *Renewable and Sustainable Energy Reviews* 47 (2015), pp. 386–398.

[24] L. Cabeza, I. Martorell, L. Miró, A. Fernández, and C. Barreneche. "Introduction to thermal energy storage (TES) systems". *Advances in Thermal Energy Storage Systems.* Ed. by L. F. Cabeza. Woodhead Publishing Series in Energy. Woodhead Publishing, 2015. Chap. 1, pp. 1–28.

[25] H. Kerskes. "Thermochemical Energy Storage". *Storing Energy – With Special Reference to Renewable Energy Sources.* Ed. by T. M. Letcher. Elsevier Inc., 2016. Chap. 17, pp. 345–372.

[26] A. Fopah Lele. "A Thermochemical Heat Storage System for Households: Thermal Transfer Coupled to Chemical Reaction Investigations". Dissertation. Lüneburg, Germany: Leuphana Universität Lüneburg, 2015.

[27] T. Yan, R. Wang, T. Li, L. Wang, and I. T. Fred. "A review of promising candidate reactions for chemical heat storage". *Renewable and Sustainable Energy Reviews* 43 (2015), pp. 13–31.

[28] P. Pardo, A. Deydiera, Z. Anxionnaz-Minvielle, S. Rougéa, M. Cabassud, and P. Cognet. "A review on high temperature thermochemical heat energy storage". *Renewable and Sustainable Energy Reviews* 32 (2014), pp. 591–610.

[29] C. Bales, P. Gantenbein, D. Jaenig, and R. Weber. *Laboratory Prototypes of Thermo-Chemical and Sorption Storage Units.* A technical report of IEA Solar Heating and Cooling programme - Task 32 - subtask B. Paris, France: International Energy Agency, 2007.

[30] N. Yu, R. Z. Wang, T. X. Li, and L. W. Wang. "Progress in Sorption Thermal Energy Storage". *Energy Solutions to Combat Global Warming*. Ed. by X. Zhang and I. Dincer. Vol. 33. Lecture Notes in Energy. Switzerland: Springer International Publishing, 2017. Chap. 28, pp. 541–572.

[31] B. Michel, N. Mazet, and P. Neveu. "Experimental investigation of an innovative thermochemical process operating with a hydrate salt and moist air for thermal storage of solar energy: Global performance". *Applied Energy* 129 (2014), pp. 177–186.

[32] N. Yu, R. Wang, and L. Wang. "Theoretical and experimental investigation of a closed sorption thermal storage prototype using LiCl/water". *Energy* 93 (2015), pp. 1523–1534.

[33] G. Li, S. Qian, H. Lee, Y. Hwang, and R. Radermacher. "Experimental investigation of energy and exergy performance of short term adsorption heat storage for residential application". *Energy* 65 (2014), pp. 675–691.

[34] D. Dicaire and F. H. Tezel. "Regeneration and efficiency characterization of hybrid adsorbent for thermal energy storage of excess and solar heat". *Renewable Energy* 36.3 (2011), pp. 986–992.

[35] B. Dawoud, E.-H. Amer, and D.-M. Gross. "Experimental investigation of an adsorptive thermal energy storage". *International Journal of Energy Research* 31.2 (2007), pp. 135–147.

[36] C. Rathgeber, E. Lävemann, and A. Hauer. "Economic top-down evaluation of the costs of energy storages–A simple economic truth in two equations". *Journal of Energy Storage* 2 (2015), pp. 43–46.

[37] B. Muster-Slawitsch, W. Weiss, H. Schnitzer, and C. Brunner. "The green brewery concept - Energy efficiency and the use of renewable energy sources in breweries". *Applied Thermal Engineering* 31.13 (2011). Selected Papers from the 13th Conference on Process Integration, Modelling and Optimisation for Energy Saving and Pollution Reduction, pp. 2123–2134.

[38] Y. Kato, F. Takahashi, A. Watanabe, and Y. Yoshizawa. "Thermal analysis of a magnesium oxide/water chemical heat pump for cogeneration". *Applied Thermal Engineering* 21.10 (2001), pp. 1067–1081.

[39] T. Nuytten, P. Moreno, D. Vanhoudt, L. Jespers, A. Sole, and L. Cabeza. "Comparative analysis of latent thermal energy storage tanks for micro-CHP systems". *Applied Thermal Engineering* 59.1–2 (2013), pp. 542–549.

[40] A. Hauer and F. Fischer. "Open Adsorption System for an Energy Efficient Dishwasher". *Chemie Ingenieur Technik* 83.1-2 (2011), pp. 61–66.

[41] A. Hauer. "Adsorption Systems for TES – Design and Demonstration Projects". *Thermal Energy Storage for Sustainable Energy Consumption.* Ed. by H. O. Paksoy. Vol. 234. NATO Science Series. Springer Netherlands, 2007. Chap. 25, pp. 409–427.

[42] C. Thim. "Domestic appliances". *Technology Guide: Principles – Applications – Trends.* Ed. by H.-J. Bullinger. Berlin, Heidelberg, Germany: Springer, 2009. Chap. 11, pp. 458–461.

[43] N. J. Hewitt, M. J. Huang, M. Anderson, and M. Quinn. "Advanced air source heat pumps for UK and European domestic buildings". *Applied Thermal Engineering* 31.17–18 (2011). SET 2010 Special Issue, pp. 3713–3719.

[44] D. Connolly, H. Lund, B. Mathiesen, S. Werner, B. Möller, U. Persson, T. Boermans, D. Trier, P. Ostergaard, and S. Nielsen. "Heat Roadmap Europe: Combining district heating with heat savings to decarbonise the EU energy system". *Energy Policy* 65 (2014), pp. 475–489.

[45] J. Heier, C. Bales, and V. Martin. "Combining thermal energy storage with buildings - a review". *Renewable and Sustainable Energy Reviews* 42 (2015), pp. 1305–1325.

[46] D. Jaehnig, R. Hausner, W. Wagner, and C. Isaksson. "Thermo-Chemical Storage for Solar Space Heating in a Single-Family House". *10th International Conference on Thermal Energy Storage, ECOSTOCK.* Stockton University , New Jersey, United States of America, 2006.

[47] A. Ristić, S. Furbo, C. Moser, H. Schranzhofer, A. Lazaro, M. Delgado, C. Peñalosa, L. Zalewski, G. Diarce, C. Alkan, S. N. Gunasekara, T. Haussmann, S. Gschwander, C. Rathgeber, H. Schmit, C. Barreneche, L. Cabeza, G. Ferrer, Y. Konuklu, H. Paksoy, H. Rammelberg, G. Munz, T. Herzog, J. Jänchen, and E. Palomo del Barrio. "Engineering and processing of PCMs, TCMs and sorption materials". *Energy Procedia* 91 (2016), pp. 207–217.

[48] A. Frazzica and A. Freni. "Adsorbent working pairs for solar thermal energy storage in buildings". *Renewable Energy* 110 (2017), pp. 87–94.

[49] D. Fehrenbach, E. Merkel, R. McKenna, U. Karl, and W. Fichtner. "On the economic potential for electric load management in the German residential heating sector - An optimising energy system model approach". *Energy* 71 (2014), pp. 263–276.

[50] K. Johannes, F. Kuznik, J.-L. Hubert, F. Durier, and C. Obrecht. "Design and characterisation of a high powered energy dense zeolite thermal energy storage system for buildings". *Applied Energy* 159 (2015), pp. 80–86.

[51] J. Xu, R. Wang, and Y. Li. "A review of available technologies for seasonal thermal energy storage". *Solar Energy* 103 (2014), pp. 610–638.

[52] B. Mette, H. Kerskes, and H. Drück. "Concepts of long-term thermochemical energy storage for solar thermal applications - Selected examples". *Energy Procedia* 30 (2012), pp. 321–330.

[53] D. Dicaire and F. H. Tezel. "Use of adsorbents for thermal energy storage of solar or excess heat: improvement of energy density". *International Journal of Energy Research* 37.9 (2013), pp. 1059–1068.

[54] M. Gaeini, H. Zondag, and C. Rindt. "Effect of kinetics on the thermal performance of a sorption heat storage reactor". *Applied Thermal Engineering* 102 (2016), pp. 520–531.

[55] A. Lass-Seyoum, D. Borozdenko, T. Friedrich, T. Langhof, and S. Mack. "Practical Test on a Closed Sorption Thermochemical Storage System with Solar Thermal Energy". *Energy Procedia* 91 (2016). Proceedings of the 4th International Conference on Solar Heating and Cooling for Buildings and Industry (SHC 2015), pp. 182–189.

[56] W. van Helden, W. Wagner, C. Krampe-Zadler, H. Kerskes, F. Bertsch, B. Mette, and J. Jänchen. "Experimental tests on a solid sorption prototype for seasonal solar thermal storage". *Eurotherm Seminar #99 Advances in Thermal Energy Storage*. Lleida, Spain, 2014.

[57] B. Michel, P. Neveu, and N. Mazet. "Comparison of closed and open thermochemical processes, for long-term thermal energy storage applications". *Energy* 72 (2014), pp. 702–716.

[58] D. Lefebvre and F. H. Tezel. "A review of energy storage technologies with a focus on adsorption thermal energy storage processes for heating applications". *Renewable and Sustainable Energy Reviews* 67 (2017), pp. 116–125.

[59] E. Worrell, L. Bernstein, J. Roy, L. Price, and J. Harnisch. "Industrial energy efficiency and climate change mitigation". *Energy Efficiency* 2 (2009), pp. 109–123.

[60] L. Miró, J. Gasia, and L. F. Cabeza. "Thermal energy storage (TES) for industrial waste heat (IWH) recovery: A review". *Applied Energy* 179 (2016), pp. 284–301.

[61] A. Krönauer, E. Lävemann, and A. Hauer. "Mobile Sorption Heat Storage in Industrial Waste Heat Recovery". *The 12th International Conference on Energy Storage, INNOSTOCK*. Lleida, Spain, 2012.

[62] A. Krönauer, E. Lävemann, S. Brückner, and A. Hauer. "Mobile Sorption Heat Storage in Industrial Waste Heat Recovery". *Energy Procedia* 73 (2015), pp. 272–280.

[63] F. Armanasco, L. P. M. Colombo, A. Lucchini, and A. Rossetti. "Techno-economic evaluation of commercial cogeneration plants for small and medium size companies in the Italian industrial and service sector". *Applied Thermal Engineering* 48 (2012), pp. 402–413.

[64] T. Nuytten, B. Claessens, K. Paredis, J. V. Bael, and D. Six. "Flexibility of a combined heat and power system with thermal energy storage for district heating". *Applied Energy* 104 (2013), pp. 583–591.

[65] D. Haeseldonckx, L. Peeters, L. Helsen, and W. D'haeseleer. "The impact of thermal storage on the operational behaviour of residential CHP facilities and the overall CO_2 emissions". *Renewable and Sustainable Energy Reviews* 11.6 (2007), pp. 1227–1243.

[66] A. Arteconi, N. Hewitt, and F. Polonara. "State of the art of thermal storage for demand-side management". *Applied Energy* 93 (2012), pp. 371–389.

[67] C. Lauterbach, B. Schmitt, U. Jordan, and K. Vajen. "The potential of solar heat for industrial processes in Germany". *Renewable and Sustainable Energy Reviews* 16.7 (2012), pp. 5121–5130.

[68] M. van der Pal, A. Wemmers, S. Smeding, and R. de Boer. "Technical and economical feasibility of the hybrid adsorption compression heat pump concept for industrial applications". *Applied Thermal Engineering* 61.2 (2013), pp. 837–840.

[69] A. Hauer. "Sorption Theory for Thermal Energy Storage". *Thermal Energy Storage for Sustainable Energy Consumption.* Ed. by H. O. Paksoy. Vol. 234. NATO Science Series. Springer Netherlands, 2007. Chap. 24, pp. 393–408.

[70] Y. Shirai and N. Osaka. "Computational Study on Energy Savings and CO2 Reduction from Combined Heat and Power with Chemical Heat Storage". *International Journal of Environmental Science and Development* 7.10 (2016), pp. 772–777.

[71] R. de Boer, S. Smeding, and P. Bach. "Heat storage systems for use in an industrial batch process: (Results of) a case study". *10th International Conference on Thermal Energy Storage, ECOSTOCK.* Stockton University , New Jersey, United States of America, 2006.

[72] B. Mette, H. Kerskes, H. Drück, and H. Müller-Steinhagen. "Experimental and numerical investigations on the water vapor adsorption isotherms and kinetics of binderless zeolite 13X". *International Journal of Heat and Mass Transfer* 71 (2014), pp. 555–561.

[73] L. Jiang, F. Zhu, L. Wang, C. Liu, and R. Wang. "Experimental investigation on a $MnCl_2$-$CaCl_2$-NH_3 thermal energy storage system". *Renewable Energy* 91 (2016), pp. 130–136.

[74] F. Zhu, L. Jiang, L. Wang, and R. Wang. "Experimental investigation on a $MnCl_2$-$CaCl_2$-NH_3 resorption system for heat and refrigeration cogeneration". *Applied Energy* 181 (2016), pp. 29–37.

[75] D. Steen, M. Stadler, G. Cardoso, M. Groissboeck, N. DeForest, and C. Marnay. "Modeling of thermal storage systems in MILP distributed energy resource models". *Applied Energy* 137 (2015), pp. 782–792.

[76] F. P. Incropera, D. P. DeWitt, T. L. Bergman, and A. S. Lavine. *Fundamentals of Heat and Mass Transfer*. 6th ed. United States of America: John Wiley & Sons, 2007.

[77] G. Zanganeh, A. Pedretti, S. Zavattoni, M. Barbato, and A. Steinfeld. "Packed-bed thermal storage for concentrated solar power – Pilot-scale demonstration and industrial-scale design". *Solar Energy* 86.10 (2012), pp. 3084–3098.

[78] R. Anderson, S. Shiri, H. Bindra, and J. F. Morris. "Experimental results and modeling of energy storage and recovery in a packed bed of alumina particles". *Applied Energy* 119 (2014), pp. 521–529.

[79] G. Angrisani, M. Canelli, C. Roselli, and M. Sasso. "Calibration and validation of a thermal energy storage model: Influence on simulation results". *Applied Thermal Engineering* 67.1-2 (2014), pp. 190–200.

[80] F. Schaube, I. Utz, A. Wörner, and H. Müller-Steinhagen. "De- and rehydration of $Ca(OH)_2$ in a reactor with direct heat transfer for thermo-chemical heat storage. Part B: Validation of model". *Chemical Engineering Research and Design* 91.5 (2013), pp. 865–873.

[81] C. Bales, D. Jaehnig, H. Kerskes, and H. Zondag. *Store Models for Chemical and Sorption Storage Units*. A technical report of IEA Solar Heating and Cooling programme - Task 32 - subtask B. Paris, France: International Energy Agency, 2008.

[82] F. Lanzerath, J. Seiler, U. Bau, and A. Bardow. "Optimal design of adsorption chillers based on a validated dynamic object-oriented model". *Science and Technology for the Built Environment* 21.3 (2015), pp. 248–257.

[83] A. Fopah Lele, F. Kuznik, O. Opel, and W. K. Ruck. "Performance analysis of a thermochemical based heat storage as an addition to cogeneration systems". *Energy Conversion and Management* 106 (2015), pp. 1327–1344.

[84] A. Mawire, M. McPherson, and R. van den Heetkamp. "Thermal performance of a small oil-in-glass tube thermal energy storage system during charging". *Energy* 34.7 (2009), pp. 838–849.

[85] S. Pal, M. R. Hajj, W. P. Wong, and I. K. Puri. "Thermal energy storage in porous materials with adsorption and desorption of moisture". *International Journal of Heat and Mass Transfer* 69 (2014), pp. 285–292.

[86] E. Osterman, K. Hagel, C. Rathgeber, V. Butala, and U. Stritih. "Parametrical analysis of latent heat and cold storage for heating and cooling of rooms". *Applied Thermal Engineering* 84 (2015), pp. 138–149.

[87] U. Bau. "Design and operation of adsorption energy systems using dynamic simulation, optimization and model predictive control". Dissertation in progress. Aachen, Germany: RWTH Aachen University, 2018.

[88] L. Yong and K. Sumathy. "Review of mathematical investigation on the closed adsorption heat pump and cooling systems". *Renewable and Sustainable Energy Reviews* 6.4 (2002), pp. 305–338.

[89] T. Nagel, S. Beckert, C. Lehmann, R. Gläser, and O. Kolditz. "Multi-physical continuum models of thermochemical heat storage and transformation in porous media and powder beds - A review". *Applied Energy* 178 (2016), pp. 323–345.

[90] M. Duquesne, J. Toutain, A. Sempey, S. Ginestet, and E. P. del Barrio. "Modeling of a nonlinear thermochemical energy storage by adsorption on zeolites". *Applied Thermal Engineering* 71.1 (2014), pp. 469–480.

[91] G. Santori, A. Sapienza, and A. Freni. "A dynamic multi-level model for adsorptive solar cooling". *Renewable Energy* 43.0 (2012), pp. 301–312.

[92] X. Wang and H. T. Chua. "Two bed silica gel-water adsorption chillers: An effectual lumped parameter model". *International Journal of Refrigeration* 30.8 (2007), pp. 1417–1426.

[93] F. Lanzerath. "Modellgestützte Entwicklung von Adsorptionswärmepumpen (in German)". Dissertation. Aachen, Germany: RWTH Aachen University, 2013.

[94] M. Fernandes, G. Brites, J. Costa, A. Gaspar, and V. Costa. "Modeling and parametric analysis of an adsorber unit for thermal energy storage". *Energy* 102 (2016), pp. 83–94.

[95] H. T. Chua, K. C. Ng, W. Wang, C. Yap, and X. L. Wang. "Transient modeling of a two-bed silica gel-water adsorption chiller". *International Journal of Heat and Mass Transfer* 47.4 (2004), pp. 659–669.

[96] G. J. V. N. Brites. "Desenvolvimento e otimização de um sistema de refrigeração solar por adsorção (in Portuguese)". PhD thesis. Coimbra, Portugal: Departamento de Engenharia Mecânica, Faculdade de Ciências e Tecnologia, Universidade de Coimbra, 2013.

[97] A. Fopah Lele, F. Kuznik, T. Osterland, and W. K. Ruck. "Thermal synthesis of a thermochemical heat storage with heat exchanger optimization". *Applied Thermal Engineering* 101 (2016), pp. 669–677.

[98] G. Gartler, D. Jähnig, G. Purkarthofer, and W. Wagner. "Development of a High Energy Density Sorption Storage System". *Proceedings of the EuroSun Conference, Freiburg, Germany.* 2004.

[99] D. Lefebvre, P. Amyot, B. Ugur, and F. H. Tezel. "Adsorption Prediction and Modeling of Thermal Energy Storage Systems: A Parametric Study". *Industrial & Engineering Chemistry Research* 55.16 (2016), pp. 4760–4772.

[100] B. Michel, N. Mazet, S. Mauran, D. Stitou, and J. Xu. "Thermochemical process for seasonal storage of solar energy: Characterization and modeling of a high density reactive bed". *Energy* 47.1 (2012). Asia-Pacific Forum on Renewable Energy 2011, pp. 553–563.

[101] P. Tatsidjodoung, N. Le Pierrès, J. Heintz, D. Lagre, L. Luo, and F. Durier. "Experimental and numerical investigations of a zeolite 13X/water reactor for solar heat storage in buildings". *Energy Conversion and Management* 108 (2016), pp. 488–500.

[102] M. Verde, L. Cortés, J. Corberán, A. Sapienza, S. Vasta, and G. Restuccia. "Modelling of an adsorption system driven by engine waste heat for truck cabin A/C. Performance estimation for a standard driving cycle". *Applied Thermal Engineering* 30.13 (2010), pp. 1511–1522.

[103] A. Pesaran, H. Lee, Y. Hwang, R. Radermacher, and H.-H. Chun. "Review article: Numerical simulation of adsorption heat pumps". *Energy* 100 (2016), pp. 310–320.

[104] C. Chen, Z. Xia, R. Wang, and J. Kiplagat. "Study on a silica gel-water adsorption chiller integrated with a closed wet cooling tower". *International Journal of Thermal Sciences* 49.3 (2010), pp. 611–620.

[105] J. Belmonte, P. Eguía, A. Molina, J. Almendros-Ibáñez, and R. Salgado. "A simplified method for modeling the thermal performance of storage tanks containing PCMs". *Applied Thermal Engineering* 95 (2016), pp. 394–410.

[106] P. Gantenbein. "Adsorption speed and mass transfer zone analysis of water vapour on the solid sorbent materials zeolite and silicagel with the focus on the heat

exchanger design". *10th International Conference on Thermal Energy Storage, ECOSTOCK*. Stockton University , New Jersey, United States of America, 2006.

[107] J. Binkert, J. Lauer, A. Diaconu, W. Russ, H. Schreiber, and A. Bardow. *Entwicklung einer Verfahrenskombination aus Zeolithwärmepumpe, Vakuumeindampfsystem und Blockheizkraftwerk zur energieeffizienten Wärmeversorgung von Brauereien (in German)*. final report AZ 23803. Bamberg, Germany: Deutscher Bund für Umwelt und Naturschutz (DBU), 2012.

[108] R. Gasper. "Entwicklung einer kompakten zweimodularen Adsorptionswärmepumpe (in German)". Dissertation. Aachen, Germany: RWTH Aachen University, 2008.

[109] M. Fernandes, G. Brites, J. Costa, A. Gaspar, and V. Costa. "Review and future trends of solar adsorption refrigeration systems". *Renewable and Sustainable Energy Reviews* 39 (2014), pp. 102–123.

[110] T. Núñez. "Charakterisierung und Bewertung von Adsorbentien für Wärmetransformationsanwendungen (in German)". Dissertation. Freiburg, Germany: Albert-Ludwigs-Universität Freiburg im Breisgau, 2001.

[111] D. Schawe. "Theoretical and Experimental Investigations of an Adsorption Heat Pump with Heat Transfer between two Adsorbers". Dissertation. Stuttgart, Germany: Universität Stuttgart, 2001.

[112] B. Sturm, S. Hugenschmidt, S. Joyce, W. Hofacker, and A. P. Roskilly. "Opportunities and barriers for efficient energy use in a medium-sized brewery". *Applied Thermal Engineering* 53.2 (2013), pp. 397–404.

[113] B. Sturm, M. Butcher, Y. Wang, Y. Huang, and T. Roskilly. "The feasibility of the sustainable energy supply from bio wastes for a small scale brewery - A case study". *Applied Thermal Engineering* 39 (2012), pp. 45–52.

[114] P. Voll, C. Klaffke, M. Hennen, and A. Bardow. "Automated superstructure-based synthesis and optimization of distributed energy supply systems". *Energy* 50 (2013), pp. 374–388.

[115] D. W. Green and R. H. Perry, eds. *Perry's Chemical Engineers' Handbook*. 8th ed. New York, United States of America: McGraw-Hill, 2008.

[116] P. Fritzson. *Principles of Object-Oriented Modeling and Simulation with Modelica 2.1*. Wiley-IEEE Press, 2004.

[117] M. Gräber, K. Kosowski, C. Richter, and W. Tegethoff. "Modelling of heat pumps with an object-oriented model library for thermodynamic systems". *Mathematical and Computer Modelling of Dynamical Systems* 16.3 (2010), pp. 195–209.

[118] U. Bau, F. Lanzerath, M. Gräber, S. Graf, H. Schreiber, N. Thielen, and A. Bardow. "Adsorption energy systems library - Modeling adsorption based chillers, heat pumps, thermal storages and desiccant systems". *Proceedings of the 10th International Modelica Conference.* Ed. by H. Tummescheit and K.-E. Årzén. Vol. 96. Linköping electronic conference proceedings. Modelica Association. Linköping, Sweden, 2014, pp. 875–883.

[119] M. M. Dubinin. "Physical Adsorption of Gases and Vapors in Micropores". *Progress in Surface and Membrane Science.* Ed. by D. A. Cadenhead, J. Danielli, and M. Rosenberg. Vol. 9. 111 Fifth Avenue, New York, United States of America: Academic Press, Inc., 1975.

[120] E. Glueckauf. "Theory of Chromatography Part 10. Formulae for Diffusion into Spheres and their Application to Chromatography". *Transactions of the Faraday Society* 51.11 (1955), pp. 1540–1551.

[121] L. Schnabel. "Experimentelle und numerische Untersuchung der Adsorptionskinetik von Wasser an Adsorbens-Metallverbundstrukturen (in German)". Dissertation. Berlin, Germany: Technische Universität Berlin, 2009.

[122] VDI-Gesellschaft Verfahrenstechnik und Chemieingenieurwesen, ed. *VDI-Wärme Atlas (in German).* 9th ed. Berlin, Germany: Springer-Verlag, 2002.

[123] H. Herwig and A. Moschallski. *Wärmeübertragung (in German).* 2nd ed. Wiesbaden, Germany: Vieweg + Teubner, 2009.

[124] I. S. Girnik and Y. I. Aristov. "Dynamics of water vapour adsorption by a monolayer of loose AQSO^{TM}-FAM-Z02 grains: Indication of inseparably coupled heat and mass transfer". *Energy* 114 (2016), pp. 767–773.

[125] MAN Engines. *Power - Gasmotoren für Blockheizkraftwerke (in German).* URL www.truck.man.eu/man/media/content_medien/doc/global_engines/power/BR_Power_Gas_DE.pdf. online, accessed 31/08/2016.

[126] E. Fabrizio. "Modelling of multi-energy systems in buildings". English. Dissertation. Lyon, France: Institut National des Sciences Appliquées de Lyon, 2008.

[127] F. Cottone and H. Mehling. *Effiziente Wärmespeicher für den Temperaturbereich 100–150 °C (in German).* final report BUT 024. Karlsruhe, Germany: Ministerium für Umwelt, Naturschutz und Verkehr Baden-Württemberg, 2009.

[128] Umweltbundesamt. *Brutto-Brennstoffnutzungsgrad fossiler Kraftwerke 2013 (in German).* URL www.umweltbundesamt.de/daten/energiebereitstellung-verbrauch/stromerzeugung. online, accessed 25/11/2013. Bundesrepublik Deutschland, 2013.

[129] H. Baehr and K. Stephan. *Wärme- und Stoffübertragung (in German)*. 8th ed. Heidelberg, Germany: Springer, 2013.

[130] F. Agyenim, N. Hewitt, P. Eames, and M. Smyth. "A review of materials, heat transfer and phase change problem formulation for latent heat thermal energy storage systems (LHTESS)". *Renewable and Sustainable Energy Reviews* 14.2 (2010), pp. 615–628.

[131] J. Gustafsson, J. Delsing, and J. van Deventer. "Improved district heating substation efficiency with a new control strategy". *Applied Energy* 87.6 (2010), pp. 1996–2004.

[132] Working Group 1 of the Joint Committee for Guides in Metrology. *Evaluation of measurement data - Guide to the expression of uncertainty in measurement*. JCGM 100:2008 (GUM 1995 with minor corrections). 2008.

[133] Deutsche Rockwool Mineralwoll GmbH & Co. OHG. *Rockwool Klimarock (in German)*. Gladbeck, Germany.

[134] S. Churchill and H. Chu. "Correlating equations for laminar and turbulent free convection from a vertical plate". *International Journal of Heat and Mass Transfer* 18 (1975), pp. 1323–1329.

[135] S. W. Churchill and R. Usagi. "A General Expression for the Correlation of Rates of Transfer and Other Phenomena". *AIChE Journal* 18.6 (1972), pp. 1121–1128.

[136] T. Fujii and H. Uehara. "Laminar natural-convective heat transfer from the outer surface of a vertical cylinder". *International Journal of Heat and Mass Transfer* 13 (1970), pp. 607–615.

[137] A. Griesinger, K. Spindler, and E. Hahne. "Measurements and theoretical modelling of the effective thermal conductivity of zeolites". *International Journal of Heat and Mass Transfer* 42.23 (1999), pp. 4363–4374.

[138] C. Bales, P. Gantenbein, D. Jaehnig, H. Kerskes, M. van Essen, and H. Z. R. Weber. "Chemical and Sorption Storage – Results from IEA-SHC Task 32". *Proceedings of the EuroSun Conference, 1st international Congress on Heating, Cooling and Buildings*. Lisbon, Portugal, 2008.

[139] Q. Li, A. Shirazi, C. Zheng, G. Rosengarten, J. A. Scott, and R. A. Taylor. "Energy concentration limits in solar thermal heating applications". *Energy* 96 (2016), pp. 253–267.

[140] M. Maivel and J. Kurnitski. "Low temperature radiator heating distribution and emission efficiency in residential buildings". *Energy and Buildings* 69 (2014), pp. 224–236.

[141] G. Zanganeh, A. Pedretti, A. Haselbacher, and A. Steinfeld. "Design of packed bed thermal energy storage systems for high-temperature industrial process heat". *Applied Energy* 137 (2015), pp. 812–822.

[142] M.-H. Simonot-Grange, F. B.-E. Hannouni, and O. Bracieux-Bouillot. "Proprietés physico-chimiques de l'eau adsorbée dans les zeolithes 13X et 4A. II. Capacités thermiques de l'eau adsorbée, du systeme zeolithe-eau et de la zeolithe anhydre (in French)". *Thermochimica Acta* 101 (1986), pp. 217–230.

[143] T. Trucano, L. Swiler, T. Igusa, W. Oberkampf, and M. Pilch. "Calibration, validation, and sensitivity analysis: What's what". *Reliability Engineering & System Safety* 91.10–11 (2006). The Fourth International Conference on Sensitivity Analysis of Model Output (SAMO 2004), pp. 1331–1357.

[144] S. Graf, F. Lanzerath, A. Sapienza, A. Frazzica, A. Freni, and A. Bardow. "Prediction of SCP and COP for adsorption heat pumps and chillers by combining the large-temperature-jump method and dynamic modeling". *Applied Thermal Engineering* 98 (2016), pp. 900–909.

[145] M. Proust. *JMP Design of Experiments Guide.* Vol. 27513. Cary, NC, United States of America: SAS Institute Inc., 2009.

[146] M. Wetter. *GenOpt Generic optimization program User Manual Version 3.1.0.* Lawrence Berkeley National Laboratory Environmental Energy Technologies Division Building Technologies Department Simulation Research Group. Berkeley, Ca, United States of America, 2011.

[147] A. Buonomano, F. Calise, and M. Vicidomini. "A dynamic model of an innovative high-temperature solar heating and cooling system". *Thermal Science* 20.4 (2016), pp. 1121–1133.

[148] J. Huang, J. Zhang, and L. Wang. "Review of vapor condensation heat and mass transfer in the presence of non-condensable gas". *Applied Thermal Engineering* 89 (2015), pp. 469–484.

[149] M. Schmidt and M. Linder. "Thermochemische Energiespeicherung zum saisonalen Ausgleich zwischen Stromangebot und Heizwärmebedarf (in German)". *Chemie Ingenieur Technik* 88.9 (2016), pp. 1267–1267.

[150] K. E. N'Tsoukpoe, T. Schmidt, H. U. Rammelberg, B. A. Watts, and W. K. Ruck. "A systematic multi-step screening of numerous salt hydrates for low temperature thermochemical energy storage". *Applied Energy* 124 (2014), pp. 1–16.

[151] M. Fernandes, G. Brites, J. Costa, A. Gaspar, and V. Costa. "A thermal energy storage system provided with an adsorption module – Dynamic modeling and viability study". *Energy Conversion and Management* 126 (2016), pp. 548–560.

[152] F. Kuznik, K. Johannes, and C. Obrecht. "Chemisorption heat storage in buildings: State-of-the-art and outlook". *Energy and Buildings* 106 (2015), pp. 183–191.

[153] M. Deutsch, D. Müller, C. Aumeyr, C. Jordan, C. Gierl-Mayer, P. Weinberger, F. Winter, and A. Werner. "Systematic search algorithm for potential thermochemical energy storage systems". *Applied Energy* 183 (2016), pp. 113–120.

[154] S. K. Henninger, S.-J. Ernst, L. Gordeeva, P. Bendix, D. Fröhlich, A. D. Grekova, L. Bonaccorsi, Y. Aristov, and J. Jänchen. "New materials for adsorption heat transformation and storage". *Renewable Energy* 110 (2017), pp. 59–68.

[155] M. Richter, M. Bouché, and M. Linder. "Heat transformation based on $CaCl_2$-H_2O – Part A: Closed operation principle". *Applied Thermal Engineering* 102 (2016), pp. 615–621.

[156] S. Mauran, H. Lahmidi, and V. Goetz. "Solar heating and cooling by a thermochemical process. First experiments of a prototype storing 60 kWh by a solid/gas reaction". *Solar Energy* 82.7 (2008), pp. 623–636.

[157] N. Yu, R. Wang, Z. Lu, L. Wang, and T. Ishugah. "Evaluation of a three-phase sorption cycle for thermal energy storage". *Energy* 30 (2014), pp. 1–11.

[158] Y. Zhao, R. Wang, T. Li, and Y. Nomura. "Investigation of a 10 kWh sorption heat storage device for effective utilization of low-grade thermal energy". *Energy* 113 (2016), pp. 739–747.

[159] O. Myagmarjav, M. Zamengo, J. Ryu, and Y. Kato. "Energy density enhancement of chemical heat storage material for magnesium oxide/water chemical heat pump". *Applied Thermal Engineering* 91 (2015), pp. 377–386.

[160] T. Nonnen, S. Beckert, K. Gleichmann, A. Brandt, B. Unger, H. Kerskes, B. Mette, S. Bonk, T. Badenhop, F. Salg, and R. Gläser. "Erprobung eines thermochemischen Langzeitwärmespeichers auf Basis eines Zeolith/Salz-Komposits (in German)". *Chemie Ingenieur Technik* 88.3 (2016), pp. 363–371.

[161] W. Lutz, H. Toufar, R. Kurzhals, and M. Suckow. "Investigation and Modeling of the Hydrothermal Stability of Technically Relevant Zeolites". *Adsorption* 11 (3 2005), pp. 405–413.

[162] J. Jänchen, K. Schumann, E. Thrun, A. Brandt, B. Unger, and U. Hellwig. "Preparation, hydrothermal stability and thermal adsorption storage properties of

binderless zeolite beads". *International Journal of Low-Carbon Technologies* 7.4 (2012), pp. 275–279.

[163] DIN German Institute for Standardization, Commission for Electrical, Electronic and Information Technologies of DIN and VDE. *Industrial platinum resistance thermometers and platinum temperature sensors IEC 60751:2008)*. 2009.

[164] DIN German Institute for Standardization, Commission for Electrical, Electronic and Information Technologies of DIN and VDE. *Thermocouples; tolerances; identical with IEC 60584-2:1982 (status of 1989)*. 1992.

[165] NIST/SEMATECH. *e-Handbook of Statistical Methods.* online, accessed 21/01/2016. Statistical Engineering Division (SED) of the Information Technology Laboratory (ITL) of the National Institute of Standards and Technology (NIST). URL: www.itl.nist.gov/div898/handbook/mpc/section3/mpc3652.htm.

Aachener Beiträge zur Technischen Thermodynamik

ABTT 1
Phillip Voll
Automated Optimization-Based Synthesis of Distributed Energy Supply Systems
1. Auflage 2014
ISBN 978-3-86130-474-6

ABTT 2
Johannes Jung
Comparative Life Cycle Assessment of Industrial Multi-Product Processes
1. Auflage 2014
ISBN 978-3-86130-471-5

ABTT 3
Franz Lanzerath
Modellgestützte Entwicklung von Adsobtionswärmepumpen
1. Auflage 2014
ISBN 978-3-86130-472-2

ABTT 4
Thorsten Brands
Einfluss der Gemischzusammensetzung auf die Verbrennung im Diesel- und GCAI-Motor
1. Auflage 2014
ISBN 978-3-95886-006-3

ABTT 5
Dominique Dechambre
Efficient Measurement of Liquid-Liquid Equilibria using Automation and Optimal Experimental Design
1. Auflage 2016
ISBN 978-395886-077-3

ABTT 6
Niklas von der Aßen
From Life-Cycle Assesement towards life-Cycle Design of Carbon Dioxide Capture and Utilization
1. Auflage 2016
ISBN 978-3-95886-080-3

ABTT 7
Matthias Lampe
Integrated Process and Organic Rankine Cycle Working Fluid Design in the Continuous-Molecular Targeting Framework
1. Auflage 2016
ISBN 978-3-95886-086-5

ABTT 8
Thomas Hülser
Optische Untersuchung der Zündvorgänge und deren Auswirkung auf die Verbrennung in PKW-Motoren

Aachener Beiträge zur Technischen Thermodynamik

ABTT 9
Malte Döntgen
Reaction Models from Reactive Molecular Dynamics and High-Level Kinetics Predictions
1. Auflage 2016
ISBN 978-3-95886-156-5

ABTT 10
Heike Schreiber
Experiments and Validated Models for Adsorption Thermal Energy Storage in Industrial and Residential Application
1. Auflage 2017
ISBN 978-3-95886-178-7